THE OXFORD BOOK OF FOOD PLANTS

THE OXFORD BOOK
OF FOOD PLANTS

Illustrations by
B. E. NICHOLSON

Text by

S. G. HARRISON
(TEMPERATE PLANTS)

G. B. MASEFIELD
(TROPICAL PLANTS)

MICHAEL WALLIS
(SOME TEMPERATE FRUITS)

OXFORD UNIVERSITY PRESS

Oxford University Press, Ely House, London W.1

GLASGOW NEW YORK TORONTO MELBOURNE WELLINGTON
CAPE TOWN IBADAN NAIROBI DAR ES SALAAM LUSAKA ADDIS ABABA
DELHI BOMBAY CALCUTTA MADRAS KARACHI LAHORE DACCA
KUALA LUMPUR SINGAPORE HONG KONG TOKYO

Printed in Great Britain by Jesse Broad Ltd., Altrincham, Cheshire.

Contents

ACKNOWLEDGEMENTS

The food plants illustrated in this book have been drawn from live specimens in the main. This was made possible by the generous and skilled collaboration of many university departments, nurserymen, societies, institutions, and individuals. Among the many to whom we owe especial gratitude are the Director and Staff of The Royal Botanic Garden, Kew; the Keeper and Staff of the University Botanic Garden, Oxford; the Welsh Plant Breeding Station, Aberystwyth; Staff of the Fairchild Tropical Garden, Miami, Florida; the Director of the Division of Tropical Research, Tela Railroad Company, La Lima, Honduras; the Director of the Botanical Garden, University of California (Berkeley); Hong Kong University Botany Department; the Professor of Botany, University of Ghana, Legon; the Southern Circle of the Botanical Survey of India; and A. Thornton Jackson who procured many specimens from Malaysia.

Nurserymen in Great Britain were generous in providing plants, fruits, and seeds, particularly Scott's Nurseries (Merriott, Somerset); Sutton & Sons (Reading); Alexander Brown (Perth); Blanchard's Nurseries (Ludwell); and the late Margaret Brownlow of the Herb Farm (Seal, Kent).

Among those who grew plants especially to provide specimens for the illustrator are the Department of Agricultural Science, University of Oxford, in whose tropical glasshouses many plants were grown from seeds obtained mainly through the enthusiastic endeavours of Oxford University Press Branches in Nigeria, India, California, Pakistan, East Asia, and Australia; members of the Shaftesbury Gardens Society; members of Guy's Marsh Borstal Institution (Dorset); Jane Gate; and A. L. Pears.

Sources of advice, information, and practical help which ranged from lending books or colour photographs to seasonal shopping in Soho include: The Royal Horticultural Society; the Director of the Botanical Gardens and National Herbarium, Melbourne; C. T. C. Tatham of the Wine Society; Thompson and Morgan Ltd.; Ambrose Dunston; Lucie Rie; and Sue Thompson.

The plan of the book was originally drawn up by Michael Wallis, who also wrote the text of the Fruit pages, 22 – 80. S. G. Harrison, Keeper of the Botany Department, National Museum of Wales (Cardiff), wrote the text of food plants of temperate regions, and Geoffrey Masefield, Lecturer in Agriculture at Oxford University, wrote both the text on tropical items and the general articles on the history, distribution, and nutritional value of food plants. Barbara Nicholson, the sole illustrator, also designed the jacket. We wish to thank the platemakers for lavishing appropriate care and skill on the 95 colour-plates.

INTRODUCTION

The purpose of this book is to provide accurate and attractive illustrations, and textual descriptions, of the plants which serve the human race for food. From these pages, the townsman can learn what kinds of plants provide the foods which appear in his local shops; the inhabitant of any country can discover the origin of those plant foods which are imported into it; and everyone can learn of the strange foods which are important to different sections of mankind in other continents and climates. The presentation is aimed at being useful for both children and adults. Special attention is given in the illustrations to those parts of each plant which are used as food. Descriptions appear opposite the illustrations to which they refer. Fuller space is given to the most important food crops, as well as an indication of the different names by which they are known in different countries. Short descriptions of some of the less important plants closely related to those which are illustrated are included in the text.

The illustrations have been made from plants carefully chosen to show the parts most involved in food production at relevant stages of maturity; the greatest care has been taken both in painting and reproduction to preserve fidelity in colour. The text descriptions aim in general at providing for each plant simple particulars of its origin, geographical distribution, and botany, and at indicating the parts used for food, and their treatment and nutritional value; in addition, other features of special interest are also mentioned for many plants. A botanical glossary is provided to explain terms which might present difficulty to the reader.

The plants have been arranged in groups, sometimes but not always botanically related, according to the kind of food they provide. Thus cereal crops come first and are followed by sugar crops, oil crops, nuts, and legumes. Later groups include, amongst others, fruits, spices, salad plants, leaf vegetables and root crops. A reader in doubt where to find a particular plant can refer to the index which includes both botanical and common names. Wild plants which provide food have been included as well as cultivated ones, and the aim has been to omit no plant which plays any significant part in nourishing the human race. The impossible goal of including every plant ever used for food has not, however, been attempted in this book. The omissions are of slight significance, but include as particular groups: the many scores of fungi which are eaten in different parts of the world; probably some dozens of minor tropical fruits which are occasionally gathered from wild plants or more rarely planted in very small numbers; and a large number of trees and herbaceous plants whose leaves are occasionally cooked and eaten in various regions of the world. Nor is it possible to include all those plants (though some are mentioned) which men will try to eat in the desperation of famine; under such circumstances the seeds of many wild grasses are used as a substitute for cereals, and the fruits, seeds, leaves, roots, and bark of a pathetically large number of plants are tasted to see if they will sustain life.

Those who are interested in piecing together from the descriptions of individual plants a more generalised picture of the relation between man and his food plants will find at the end of the book three short articles which provide at least guidelines for further thought. These deal respectively with the domestication of food plants and their botanical affiliations, with the spread of food plants around the world in prehistoric and historic times, and with the nutritional value of food plants in human diets. Such short treatments of these large and fascinating subjects cannot be anything more than introductory, but this book will have served its purpose if, besides offering elementary information, it provides some readers also with a stimulus to further study.

GLOSSARY

Achene. A small, dry, 1-seeded, indehiscent fruit (Fig. 2).

Alternate. Arranged spirally or alternately, not in whorls or opposite pairs (Fig. 3).

Anther. Pollen-bearing part of a stamen (Fig. 2).

Appressed. Pressed against another organ but not united with it.

Aril. An extra covering to the seed in some plants.

Awn. A bristle-like projection from the tip or back of the glumes in some grasses (Text Fig., p. 2).

Axillary. In the axil, the angle between the stem and the leaf-stalk (Fig. 3).

Bipinnate. Double pinnate; with the primary divisions again divided (Fig. 4).

Bract. A modified leaf beneath a flower or part of an inflorescence (Fig. 3).

Bracteole. A small or secondary bract.

Calyx. The outer whorl of floral parts (sepals) which may be free or united (Fig. 2).

Campanulate. Bell-shaped.

Carpel. Unit of an ovary or fruit.

Caryopsis. Grain or grass-fruit, in which the seed-coat is united with the ovary wall.

Corm. The base of a stem swollen with reserve materials into a bulbous shape.

Corolla. The second or inner whorl of a flower, consisting of free or united petals (Fig. 2).

Corymb. A broad, flattish inflorescence in which the outer flowers open first (Fig. 1).

Crenate. With blunt teeth (Fig. 3).

Crenulate. With small, blunt teeth.

Culm. The flowering stem of a grass.

Cupule. A little cup.

Cyme. A repeatedly branching inflorescence, with the oldest flower at the end of each branch (Fig. 1).

Decumbent. Lying on the ground but with the ends curving upwards.

Decurrent. Extending downwards below the point of attachment.

Dichotomous. Equally forked (Fig. 1).

Disc Floret. A flower in the centre of the flower-head (of the Daisy family) (Fig. 2).

Dorsal. On the back or outer face.

Drupe. A fleshy, indehiscent fruit with a stone usually containing 1 seed (e.g. a plum) (Fig. 2).

Endosperm. Part of a seed containing most of the reserves.

Fascicle. A compact cluster.

Floret. An individual flower in a dense inflorescence, as in the Daisy and Grass families.

Glabrous. Not hairy.

Glume. Basal bracts in grass spikelets (Text Fig., p. 2).

Indehiscent. Not opening along any definite lines to shed its seeds.

Inflorescence. The flowering region or mode of flowering of a plant.

Involucre. A number of free or united bracts, surrounding or just below one or more flowers or fruits (Fig. 3).

Keel. Lower petal or fused petals ridged like the keel of a boat, as in the pea family (Fig. 3).

Lanceolate. Lance-shaped; narrow and tapering towards the tip (Fig. 4).

Node. A joint on the stem where a leaf is (or was) attached (Fig. 3).

Oblanceolate. Inverted lanceolate, the broadest part above the middle (Fig. 4).

Ovary. The female part of a flower, enclosing the ovules (Fig. 2).

Palmate. With three or more lobes or leaflets radiating like fingers from the palm of a hand (Fig. 4).

Panicle. A branched inflorescence (Fig. 1).

Papilionaceous. Type of flower characteristic of the pea family (Fig. 3).

Pappus. Modified calyx of the Compositae commonly either membranous or in the form of a 'parachute' of hairs (e.g. Dandelion) (Fig. 2).

Perianth. The sepals and petals of a flower (Fig. 2).

Petiole. Leaf-stalk (Fig. 3).

Pinnate. A compound leaf with more than 3 leaflets arranged in 2 rows on a single common stalk (Fig. 4).

Pinnatifid. Pinnately lobed, but completely divided into leaflets (Fig. 4).

Pubescent. Covered with short, soft hairs.

Raceme. An unbranched inflorescence with the individual flowers stalked (Fig. 1).

Rachis. Main axis of an inflorescence.

Receptacle. The end of the flower-stalk on which the parts of the flower are borne (Fig. 2).

Rhizome. A more or less swollen stem, wholly or partially underground.

Scabrous. Rough to the touch.

Septum. A partition.

Serrulate. With minute, forward-pointing teeth (Fig. 3).

Sessile. Not stalked.

Silicula. Like a siliqua, but short and broad in proportion to its length.

Siliqua. A fruit characteristic of the Wallflower family, elongated and pod-like, but with a central partition and opening from below by 2 valves (Fig. 2).

Spadix. A spike bearing flowers sometimes sunken, enclosed in a spathe (Fig. 1).

Spathe. Large, sheathing bract (Fig. 1).

Spike. An unbranched, elongated flower-head, bearing stalkless flowers (Fig. 1).

Spikelet. Unit of a grass flower-head, usually with 2 glumes and 1 or more florets (Text Fig., p. 2).

Spore. Minute reproductive body of a non-flowering plant.

Stamen. Male (pollen-bearing) part of flower (Fig. 2).

Standard. The broad, upper petal of a flower of the pea family (Fig. 3).

Stigma. The part of the female organ of the flower which receives the pollen (Fig. 2).

Stipule. A scaly or leaf-like outgrowth at the base of the petiole (Fig. 3).

Style. The connecting portion between stigma and ovary (Fig. 2).

Syncarp. A multiple fruit made up of small fruits united together (Fig. 2).

Ternate. Divided or arranged in threes (Fig. 4).

Tuber. Swollen, underground part of a stem or root.

Umbel. An inflorescence with branches radiating like the ribs of an umbrella (Fig. 1).

Valve. Segment of a dehiscent fruit.

Wings. Lateral petals characteristic of flowers of the pea family (Fig. 3).

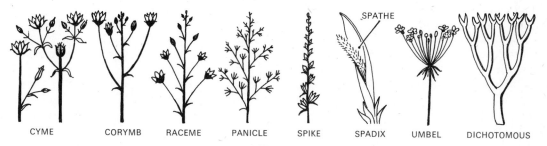

CYME   CORYMB   RACEME   PANICLE   SPIKE   SPADIX   UMBEL   DICHOTOMOUS

SPATHE

Fig. I

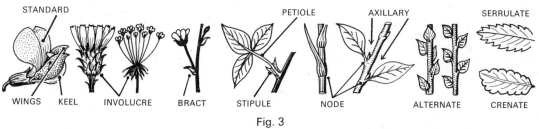

PERIANTH

COROLLA   DISC FLORETS   STYLE   STIGMA   ACHENE

CALYX   PAPPUS   ANTHER   STAMEN   OVARY   RECEPTACLE   SYNCARP   SILIQUA   DRUPE

Fig. 2

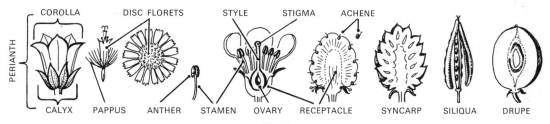

STANDARD   PETIOLE   AXILLARY   SERRULATE

WINGS   KEEL   INVOLUCRE   BRACT   STIPULE   NODE   ALTERNATE   CRENATE

Fig. 3

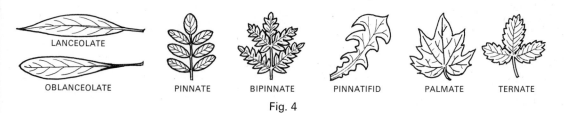

LANCEOLATE

OBLANCEOLATE   PINNATE   BIPINNATE   PINNATIFID   PALMATE   TERNATE

Fig. 4

GRAIN CROPS: WHEAT

Grain crops are by far the most important sources of plant food for man, both directly as cereals, and indirectly as fodder for domesticated livestock providing meat and dairy products. Grains are the storehouses of food for the young plant, and consist mainly of starch with some protein, and traces of minerals and vitamins. Except for a very small number of so-called pseudo-cereals (such as Buckwheat and Quinoa), all the grain crops or cereals belong to the Grass family (Gramineae). Their 'grain' is the characteristic fruit of the Grass family, known botanically as a caryopsis, in which the seed coat is fused with the ovary wall. The origins of many of the grains are uncertain; but their wild ancestors, now often uncertain, were probably the first crops that Stone Age man brought into cultivation.

The wheats are the most important cereals in temperate climates. They provide such staple food as bread, baked from flour, the fine powder which is produced by grinding wheat grains and separating out the chaff. The wheats probably evolved from wild species of *Triticum* and the related genus *Aegilops* in south-western Asia and eastern Mediterranean regions.

1-2 Bread Wheat (*Triticum aestivum*, syn. *T. vulgare*). The bread wheats are the most widely grown and the most economically important group of wheats. They are the source of the highest quality bread flours and some of their cultivated varieties produce the most suitable flour for making biscuits, cakes, and pastry. They show a great range of variation both in form and physiology. There are awned and awnless forms; spring wheats (sown in spring and harvested in late summer); winter wheats (sown in autumn and harvested in early summer); wheats with grains of whitish, amber, reddish, or even purple or bluish colour; hard wheats with grains rich in proteins, and soft wheats, mealy in texture and richer in starch. The stems are usually hollow and the spike long in proportion to its width. The glumes are usually keeled in the upper half only. Bread wheat is grown in many lands, wherever the climate is suitable, but particularly in the north and south temperate zones. It is the common wheat in most of the great wheat-growing areas of Europe, the U.S.S.R., Asia, America, and Australia, and also in the United Kingdom, which, however, has to import four-fifths of the wheat it uses. Unknown in the wild state, bread wheat is probably of complex origin through hybridisation and gene mutation, resulting in a vast hybrid group of thousands of varieties. Research workers are constantly trying to produce higher yielding varieties with resistance to rust and other diseases.

3 Durum Wheat (*Triticum durum*). The flour milled from durum or macaroni wheat is the best for making pasta (macaroni, spaghetti, etc.), because it contains a high proportion of gluten, the component of flour which becomes sticky and elastic when wetted. Most forms of durum are awned, spring or semi-winter wheats, with white, amber, red, or rarely purple grains which are usually long, hard, and translucent. The stems are either solid or thick-walled, the spikes rather stout in proportion to their length, and the outer glume distinctly keeled from base to apex. Durum wheat is second in importance to bread wheat. It is grown throughout the Mediterranean region, and in the U.S.S.R., Asia, and North and South America, particularly in arid regions.

4 Emmer (*Triticum dicoccum*), an important crop in early historic times, is still grown in a few countries, mainly for fodder.

5 Rivet, Cone, or English Wheat (*Triticum turgidum*). This is a tall, vigorous wheat, with stout heads which are sometimes branched. The stems are solid (filled with pith) in the upper part, or thick-walled. The glumes are keeled from base to apex. It was at one time the principal wheat grown in southern England, but because its flour is of low quality for breadmaking, it has largely been supplanted by bread wheat. It is still grown in some countries in Western and Southern Europe and elsewhere, mainly as food for livestock.

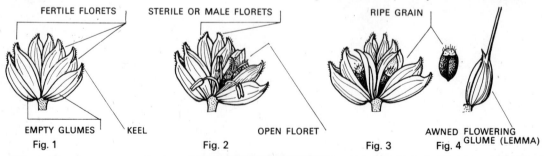

FERTILE FLORETS   STERILE OR MALE FLORETS   RIPE GRAIN

EMPTY GLUMES   KEEL   OPEN FLORET   AWNED FLOWERING GLUME (LEMMA)

Fig. 1   Fig. 2   Fig. 3   Fig. 4

Wheat spikelets (diagrammatic).
Fig. 1. Complete spikelet.   Fig. 2. Complete spikelet, with one fertile floret exposed, showing ovary and stamens.
Fig. 3. Mature spikelet with ripe grains.   Fig. 4. A single-awned flower.

LIFE SIZE   WHEAT PLANT $\times \frac{1}{8}$

1 & 2 BREAD WHEAT awnless and awned forms   1A Spikelet and grains
1B Flowering spikelet   1C Wheat plant
3 DURUM WHEAT   3A Grains   4 EMMER   4A Spikelet and grains
5 RIVET or ENGLISH WHEAT   5A Grains

3

GRAIN CROPS: RYE, OAT, BARLEY

1 Rye (*Secale cereale*). Rye is an important crop in the colder parts of north and central Europe and Russia, grown as far north as the Arctic Circle, and in mountainous areas 14,000 feet above sea level. It is a relative newcomer to cultivation, found in Iron Age sites in Britain and elsewhere, but not in earlier Egyptian or Lake Dwelling remains. Rye is reputed to have originated in Asia Minor, perhaps first occurring as a cereal field weed and then becoming accepted as a crop in its own right. In parts of Central Asia today, it still occurs wild as a weed in crops of winter wheat. Rye can be grown in cold regions and on poor soils where most other cereals would not be productive. The grain is similar in composition to wheat, and in Europe it is used chiefly for making black bread (*schwartzbrot*) which has a rather strong flavour. Scandinavian rye crisp-bread has wide popularity, because of its flavour, its value as a substitute for ordinary bread in low-calorie diets, and the fact that it will keep indefinitely. Rye is used for making whisky in America, gin in Holland, and beer in Russia. Young plants are used as fodder for livestock. The mature straw is too tough for that purpose, but it has its uses for bedding, thatching, papermaking and straw hats. Ergot (*Claviceps purpurea*), a fungus parasitic on rye, is poisonous to man and animals. Eating rye bread contaminated with ergot may cause gangrene, abortion, hallucinations, or other unpleasant symptoms. Ergot preparations are used in pharmacy.

In appearance, rye is similar to wheat but has narrow, 1-leaved glumes. The spikelets are usually 2-flowered; the lemmas are narrow, with stiff hairs on the keels, long awned.

2 Oat (*Avena sativa*). Believed to have been derived from *Avena fatua*, one of the wild oats, the common oat is known to have existed in Europe during the Bronze Age. Like rye, it may have occurred as a weed amongst other cereals before being cultivated as a separate crop. It does not seem to have become established in Britain until the Iron Age. The proportion of the oat crop used as human food is comparatively small, though porridge, grits, and similar preparations of oatmeal or rolled oats are well-known as breakfast foods in various countries. Oats are mainly used for feeding livestock, not only as ripened grain but also in the green stage as fodder, hay, or processed into pellets. Oat flour has antioxidant properties and is therefore mixed with other food-stuffs, such as margarine and ready-mixes for cakes, to help delay deterioration. In the process of milling the grain, thousands of tons of oat husks are removed yearly. Formerly, they were used as fuel or as packing material, but nowadays they have greater value as a raw material from which furfural is made. Furfural is a fluid which, with its derivatives, has many uses; for example, in making nylon, in oil refining, in synthetic rubber production, and in the manufacture of antiseptics.

Oats are easily distinguishable from other temperate cereals by the open, spreading panicle, bearing large, pendulous spikelets. Black or Bristle Oat, *Avena strigosa*, known in Wales as Ceirch Llwyd or Blewgeirch, is grown in hilly districts where conditions are unsuitable for the more usual *A. sativa*. It is distinguished from the latter by the presence of two bristles at the tip of the lemma, not to be confused with the much longer awn which is present in many cultivars of *A. sativa* as well as in *A. strigosa*.

3 Two-rowed Barley (*Hordeum distichon*). Barley bears its spikelets in threes at each node, not singly as do wheat and rye. Most of the modern cultivars commonly grown in the British Isles are two-rowed barleys, in which only the middle spikelet of each three produces seed, the other two being sterile or male. The majority of British barleys of malting quality belong to this species. 'Malting' is a process in which the grain is germinated and the very young seedlings are then dried to produce malt for brewing beer. Malt contains enzymes which convert starch to fermentable sugars. A by-product of brewing is yeast, which is used in baking and for the production of vitamin-rich yeast extracts. Pearl barley, used in soups and stews, is made by de-husking and grinding the grain. Patent barley is barley meal.

Barley of too low a standard for malting is used for feeding livestock, as are many cultivars of this and the next species which have been selected specifically for that purpose.

Two-rowed barley appears to have originated in historical times, perhaps by hybridisation between a cultivar of six-rowed barley and *Hordeum spontaneum*, a wild two-rowed barley.

4 Six-rowed Barley (*Hordeum vulgare*). This species differs from the preceding one in having three fully fertile spikelets at each node. It is little grown in Britain, but is much more widely cultivated in colder regions where hardiness is more important. Six-rowed barley is an older crop than its relative. Carbonised remains dating back to the Neolithic and Bronze Ages suggest that it was the dominant cereal in Europe throughout those times. Impressions of barley on Greek and Roman coins, about 500 B.C., are always of this species. It is used in the same ways as two-rowed barley.

Two-rowed Barley and Six-rowed Barley.

LIFE SIZE   *PLANTS* $\times \frac{1}{8}$

1 RYE   1A Spikelet and grains   1B Plant
2 OAT   2A Spikelet and grains   2B Plant
3 TWO-ROWED BARLEY   3A Plant   4 SIX-ROWED BARLEY   4A Spikelet and grains

GRAIN CROPS: MAIZE OR CORN

The only cereal crop that has an American origin is now a principal cereal crop of tropical and sub-tropical regions throughout the world. It is also known as 'Indian corn', in America simply as 'corn', and in South and Central Africa as 'mealies'.

1 Maize (*Zea mays*) is an annual crop planted from seed, and it takes 3 – 5 months to mature. The male flowers form the 'tassel' at the top of the stem; the female flowers are borne on the ears which are carried lower down the stem. The ear is covered by modified leaves which form the 'husk'. Between these leaves at the tip of the ear there protrudes the 'silk' composed of the long thin styles of the female flowers which will receive the pollen from male flowers.

The ear consists of a central structure, the 'corn cob', on which the grains are set in rows. Usually, only one or two maize ears mature on each plant. The harvesters break the maize ears off the plant and remove the leaves of the husk, usually by hand. In large scale cultivation, however, special harvesting machines remove and strip the ears. Later the grain may be removed from the cob by hand or mechanically — a process known as 'shelling'.

2-4 Maize types. There are many different types of maize grain. In the 'dent' types (2 and 2A) there is a depression caused by shrinkage at the tip of the seed. In 'flint' types (3) the grains are harder and somewhat translucent. 'Flour' types have soft grains, 'pop corn' has hard ones which burst upon heating, and 'sweet corn' has a more sugary taste. Maize grains also vary in colour, and since the male parent can influence the colour of individual seeds, one ear of a plant which has been cross-pollinated can show grains of different colours. Where maize is grown mainly for human consumption, the white meal which is produced from white-seeded types is generally preferred, but yellow maize is also common, and there are kinds whose seeds are red, purple, or almost black.

Much maize is nowadays sown from 'hybrid' seed whose production, first developed in the United States, is one of the triumphs of agricultural science and has raised yields to levels previously unattainable.

History. Maize was originally an American crop, but was brought to Europe by Columbus and has since been widely dispersed to all parts of the world. It is grown from the equator up through the latitudes where the summer is still long enough for the seed to ripen. Selected quick-maturing varieties with short stems will just ripen their seed in southern Britain, but this area is marginal for a seed crop. For use as silage, however, taller forms are grown and harvested without ripening their grain.

Uses. Maize is grown as food for both people and animals. In the United States, which is the world's largest producer, most of the crop is used for feeding livestock, and is particularly important in pig production. Argentina exports large quantities for use as an animal foodstuff.

As food for humans, maize is most important in South America and in southern and eastern Africa, where it is often the staple food; and it is also an important article of diet in south-eastern Europe. The grain is ground into a meal, usually between rollers, but in Africa often by pounding with a pestle and mortar. Commercial 'cornflour' is a finely ground form of maize meal. In Latin America, the maize meal is cooked into cakes known as *tortillas*; in Africa it is boiled with water into a gruel or porridge, rather like the Italian *polenta*. North American dishes cooked from maize products include corn bread, corn pone, and hominy. Maize, particularly sweet corn types, is sometimes cooked and eaten 'on the cob'. Slightly immature ears (4) may be picked for this purpose, as the grains are softer. The familiar breakfast food 'corn flakes' is produced from maize grain by a series of processes which include pre-cooking, flaking, and toasting. Maize grain is sometimes processed as a source of starch, and of sugar; and maize oil is increasingly important as a cooking oil.

Food Value. Maize is a good source of starch in the diet but its protein is of lower nutritional value than that of other cereals. Yellow maize contains the pigment carotene from which the human body can produce vitamin A, of which serious deficiency is found in some tropical populations. On the other hand, the disease of pellagra, caused by deficiency of nicotinic acid, is characteristically found among populations whose diet is based too exclusively on maize, but this deficiency disease is fortunately becoming rarer in the world.

PLANT × ⅛   EARS × ⅔   GRAIN AND FLOWER DETAILS × 1

1 MAIZE or CORN plant   1A Male flowers detail   1B Female flowers detail
2 DENT TYPE MAIZE ear   2A Grains   3 FLINT TYPE MAIZE grains
4 SWEET CORN MAIZE immature ear with husks   4A Grains

GRAIN CROPS: RICE

Rice is one of the world's two most important food crops (the other being wheat). As rice often supplies a very high proportion of the total food intake in the countries where it is grown, its nutritional quality is very important.

1 Rice (*Oryza sativa*) originated in Asia; it was already a staple food in China in 2800 B.C., and in India almost as early. Rice dominates the agriculture of many Asian countries, where the bulk of the world rice is produced, and is grown from the equator to as far north as Japan. Its acreage in the other continents, too, is increasing steadily — in Africa and South America largely by small farmers, and in the United States and Australia by commercial rice-growing. In Europe, where Italy has for long been the leading producer, rice growing has recently been pushed further north, especially in southern France and in Hungary.

Rice is grown on one of two systems — either in standing water, or on dry land like any other cereal. The latter form, often known as 'upland' rice, accounts for only about 10 per cent of rice acreage. Rice is adapted to growing in water by its hollow stem, which permits oxygen to pass downward and reach the roots in the flooded soil. Water may be provided by irrigation, or, where possible, by impounding rainwater in plots enclosed by low earth ridges. In some countries seed is sown direct into the field; in the United States pre-germinated seed is sometimes dropped by aeroplanes into standing water. But in Asia seed is usually sown in nurseries and the seedlings later transplanted into fields with very shallow water over the soil; the water-level is raised if possible for the main period of growth, but drained a few weeks before harvest to allow the ears to dry off and the grain to ripen. For the crop to reach maturity takes about 4 weeks in the nursery followed by 4 months in the field.

The rice plants grown in the tropics and those grown in the warm temperate zone belong to different groups of varieties and are not directly interchangeable, though hybrids have been made to combine some of the useful qualities of both. Most of the temperate rices have ears with awned grains (4) and most of the tropical rices are without awns (3).

When the crop is ripe, it is cut by hand by most tropical growers, or cut and threshed by combine harvesters where production is mechanised. In the former case, threshing is done by hand, often by beating the sheaves on a table of bamboo slats, or in small threshing machines, sometimes operated by foot pedals.

In the next process, milling, the outer husk is first removed; this exposes the grain covered by an outer brownish layer which forms the bran fraction. (In West Africa a different species, *Oryza glaberrima*, is sometimes cultivated and is called 'red rice' because its bran layer is red.) Rice is usually further milled in an operation called 'pearling' to remove the bran (which is a valuable stock-feed). Pearling leaves a white grain, sometimes further 'polished' to give it an attractive sheen. In popular parlance, however, 'pearled' rice is often referred to as 'polished'.

Nutritionally, rice is an excellent food, especially as a source of starch, but its protein content is slightly lower than in most cereals. It may be conveniently cooked by simple boiling and does not need to be ground, as most cereals do. Rice with a long grain (5) generally fetches the highest price, but most rice has shorter, rounder grains. A special type is 'glutinous' rice, which becomes sweet and sticky when cooked, and is in demand in some countries of the Far East. In parts of Asia where the diet consists too exclusively of polished, or more correctly pearled, rice, a disease called 'beri-beri' used to be prevalent. It is due to a deficiency of vitamin B_1 (thiamine) which is present in the bran fraction and embryo of rice but not in the centre of the grain. Rice-eaters can avoid beri-beri if they eat more lightly milled rice, or rice parboiled before milling. (Parboiling consists of soaking and heating the grains to a temperature below boiling point, which causes some thiamine to diffuse through the whole grain.)

Rice, boiled or fried, is the basic dish of many famous cuisines. In Southern China it is accompanied by dishes of pork, chicken, fish, and vegetables. In India it is served with curried dishes containing a variety of spiced meats and vegetables. Famous European rice dishes include Italian *risotto* and Spanish *paella*. Rice wine, known as *saki* in Japan, is prepared from the grain in some Asian countries.

Many Asian countries do not grow enough rice to supply their needs but other Asian countries, especially Thailand and Burma, are exporters. Now, however, the United States is the largest rice exporter.

American Wild Rice (*Zizania aquatica*). Also known as 'Indian rice' and 'Tuscarora rice', this annual aquatic grass is native to eastern North America, where it is valued as a food for waterfowl. The Indians harvest the grain both for their own use and for sale. When harvested, the wild rice is either sun-dried, or 'parched' by heating in a metal drum or trough over a fire. It may be eaten boiled or steamed, in soups and stews, or served with meat or game, or as a pudding. The Americans also 'pop' it in deep oil. Wild rice is reputed to have both a higher protein content than most other cereals and comparable quantities of vitamins, but it is difficult to harvest and therefore expensive.

This is a robust and ornamental grass, about 12 feet high, with a large panicle, up to 2 feet long, bearing pendulous male spikelets on its spreading or ascending lower branches and appressed female spikelets on the erect upper branches. The female spikelet is large, about 1 inch long, narrow cylindrical, with a long awn.

PLANT × ¼   *PANICLES LIFE SIZE*   *DETAILS × 2*

1 RICE plant   2 Flowering spikelet detail
3 Panicle of ripe grain   4 Panicle of awned variety
5 Details of spikelets and polished grains

9

GRAIN CROPS: TROPICAL MILLETS

A number of species of millet, the three most important and widely-grown kinds of which are illustrated, are cultivated in the tropics. The general characteristic of millets is to be small-grained cereals — but sorghum is exceptional in having larger seeds than the others and some people do not classify it as a millet. Another characteristic of the tropical millets is that they are all to a greater or lesser degree drought-resistant, and often one of them is the last crop grown where cultivation peters out into semi-desert conditions. They tolerate poor soils and (apart from sorghum) have excellent storage properties, but because of the conditions in which they are grown yields are often low. They are eaten in the form of a porridge cooked from the meal produced by grinding or pounding the grain; they can also all be used for making beer, though sorghum is the most commonly used for this purpose. Nutritionally, several of the tropical millets have a content of useful minerals which is high compared with other cereals, and finger millet is outstanding in this respect.

1-2 Sorghum (*Sorghum vulgare*) is also known in English by a variety of other names, such as 'great millet', 'kaffir corn' (in South Africa), and 'guinea corn' (in West Africa). The plants may be from 3 to 15 feet high — the illustration (1) shows only the upper part of the stem. Before flowering, the plants are quite similar to maize in general appearance, but can be differentiated from maize by the narrower leaves, the waxy bloom covering the leaves and stem, and the better-developed root system; these are all characters that enable it to thrive under drier conditions than maize. The grains are borne in a terminal ear at the top of the stem. A typical white-grained variety (1) has grains which produce a white meal, and are preferred for eating purposes. The grain of red-grained types (2) has a bitter taste and is mainly used for making beer. There are also kinds of sweet sorghum whose stems are crushed, in a similar way to sugarcane, to obtain a syrup which is produced mainly in the United States and used in cooking. Sorghum is probably of African origin, but has also been cultivated in Asia since very ancient times. It is now widely distributed in semi-arid tropical and sub-tropical regions; the most important areas of cultivation for food are in tropical Africa, central and northern India, and China. While very important as a food crop in these areas, sorghum enters little into world trade, and when it does it is mainly for use in compounding livestock feeds. The sorghum grain produced in the United States and Australia is also for livestock feeding rather than human food.

3-4 Finger Millet (*Eleusine coracana*) is so-called because the ear consists of about 5 spikes which radiate, sometimes in a curving manner, from a central point, rather as the fingers are attached to the palm of the hand. Our illustrations include an Indian variety with rather long spikes (3) and an African one with shorter more curving spikes (4). Finger millet is a short-stemmed millet, growing to about the same height as wheat or barley. The crop, which probably originated in India, is cultivated on a large scale in the drier parts of south India and Ceylon. It is also an important food crop in semi-arid areas of Africa from Rhodesia northward to the Sudan, and is the staple food of some populations in this region. It has the best storage properties of all the millets, and when stored on the farm as unthreshed heads can be kept for as much as 5 years. It has therefore a very important use as a famine reserve food over wide areas of the dry tropics, but it enters very little into trade outside the areas where it is grown. It is often planted as the first crop where bush has been newly burned and cleared, and appears to give a particularly good response to manuring with ashes in this way.

5 Bulrush Millet (*Pennisetum typhoideum*) is also, in its white-seeded forms, sometimes called 'pearl millet'. The plant is tall, of the same order of height as maize and sorghum and quite closely resembling them while in the vegetative phase. It bears at the top of the stem a long cylindrical ear which in general appearance rather resembles a bulrush and gives the crop its common name. A brown-seeded variety is illustrated (5) with part of the stem left out at the bottom, and details of its ear and spikelet are shown in 5B and 5C. This is one of the most drought-resistant millets, and is possibly more widely grown than any other crop in tropical areas of very low rainfall. It probably originated in Africa, where it provides a very important food supply in the Sudan, the far north of Nigeria, and other countries along the southern fringe of the Sahara desert. It is also an important food crop in some of the driest areas of India and Pakistan.

Other Tropical Millets. There are several other tropical millets, which are of only minor or local importance. The chief of these are teff (*Eragrostis abyssinica*) which is the most widely-grown cereal in Ethiopia but is little cultivated elsewhere, and *Digitaria exilis*, sometimes known as 'hungry rice' which is of local importance in some rather dry areas of West Africa such as the Jos plateau of Nigeria. Another very localised tropical cereal, which can hardly be classified as a millet because of its very large seeds, is *Coix lachryma-jobi* ('adlay' or 'Job's tears') which is grown in the Philippines and some other areas of south-east Asia.

PLANTS × ⅛   DETAILS LIFE SIZE   SPIKELETS × 3

1 SORGHUM white-grained type   1A Detail of ear and seed   1B Spikelets

2 RED-GRAINED SORGHUM   2A Detail of ear and seed

3 FINGER MILLET (INDIAN)   3A Spike and seed details   3B Spikelet   4 FINGER MILLET (AFRICAN)

5 BULRUSH MILLET   5A Detail of ear and seed   5B Spikelets

11

GRAIN CROPS: TEMPERATE MILLETS

1 Common Millet (*Panicum miliaceum*). This grain is widely grown as a food crop of man and livestock, especially in Russia, China, Japan, India, Southern Europe, and parts of North America, under numerous names, such as 'Proso', 'Indian', 'Broom-corn' and 'Hog' millet. It has been cultivated since prehistoric times in southern Europe and in Asia. It was the 'milium' of the Romans, and one of the millets of the Old Testament; in Ezekiel 4:9, there is a reference to its use for making bread. The grain is nutritious, containing not only carbohydrates but also 10 per cent protein and 4 per cent fat. It is used also as a food for livestock, and in Russia as a source of starch. In Britain, its principal use is as birdseed, and plants found on rubbish tips, especially in urban districts, probably often originate from seed discarded with the cleanings from cages.

Common Millet is an annual, up to 3 – 4 feet high, with numerous, long, ascending branches forming a rather compact panicle, often nodding at the tip, its shape suggesting an old-fashioned broom (hence one of its vernacular names). The smooth, glossy grains are ovoid, about $\frac{1}{8}$ inch long, and may be whitish, straw-coloured, or reddish-brown.

2 Little Millet (*Panicum miliare*) is smaller in all its parts than *Panicum miliaceum* but otherwise is generally similar to it and is capable on a good soil of producing about the same yield. However, its chief virtue is its capacity to produce a moderate yield even on very poor soils and to withstand both drought and water-logging better than most crops. It is chiefly grown in India both just inside and just outside the tropic of Cancer, and although only a minor crop on the national scale, it is most important in the states of Madhya Pradesh and Uttar Pradesh.

3 Foxtail Millet (*Setaria italica*) is unknown in the wild state, except as a weed escaped from cultivation. Despite the specific epithet '*italica*' and various vernacular names referring to European countries — e.g. 'Italian', 'German', and 'Hungarian' millet — *Setaria italica* probably originated in Asia and may have been derived from *Setaria viridis*, a common weed, widespread in the Old World. It was a sacred plant in China as early as 2700 B.C. and was also known to the Lake Dwellers in Europe. It is cultivated in all sub-tropical and warm temperate countries, being used for human food wherever it is grown, except in North America. It is an important hay crop and is also suitable for silage. In Russia, it is used for brewing beer. In Britain it is, like Common Millet, best known as birdseed.

Foxtail Millet is a robust annual, often 3 – 5 feet high. Its inflorescence is a dense, spike-like panicle, up to 12 inches long. The spikelets are surrounded by bristles which remain on the rachis after the upper floret ('seed') has been shed. The grain is about $\frac{1}{16}$ inch long. It is very variable in colour: white, yellow, red, brown, or black.

4 Japanese Millet (*Echinochloa frumentacea*) is widely cultivated in warm regions, for use as food or forage. In Japan and Korea the grain is ground into meal and made into a kind of porridge. In America it is regarded primarily as a forage crop and is not used for human food. In Britain plants occasionally appear on town rubbish tips.

Japanese Millet grows up to 4 feet high, branching from the base. The leaves, about $\frac{1}{2}$ inch wide, have sharp, minutely toothed margins. The inflorescence is up to 6 inches long, densely branched and usually purple-tinged, with awnless scabrous spikelets. The grain is light brown to purple.

Several other millets, cultivated in various parts of the world for food and forage, are occasionally seen in the British Isles. 'Barnyard Millet' (*Echinochloa crus-galli*), which is closely related to Japanese Millet, sometimes becomes more or less naturalised in waste places and cultivated ground, producing seed in favourable seasons.

PLANTS × ⅛   EARS AND SEEDS LIFE SIZE   SPIKELETS × 3

1 COMMON MILLET plant	1A Ripe ear and seed	1B Spikelets
2 LITTLE MILLET plant	2A Ripe ear and seed	2B Spikelets
3 FOXTAIL MILLET plant	3A Ripe ear and seed	3B Spikelets
4 JAPANESE MILLET plant	4A Ripe ear and seed	4B Spikelets

SUGAR CROPS

Although sugar is manufactured by all green plants, and may be stored in roots, flowers, bulbs, and fruits, the plants from which sugar is refined for commercial use are pre-eminently sugar cane and sugar beet. Sugar is an energy-producing food which man can assimilate very quickly.

1 **Sugar Cane** (*Saccharum officinarum*) provides more than half the world's supply of sugar. It gives its best yields in the tropics, but is also grown in the sub-tropics as far north as Louisiana in the United States and southern Spain, and as far south as New South Wales. The crop is planted by stem cuttings or 'setts', and about a year later in tropical lowlands the first harvest of stems is ready to be taken. Cutting is still most commonly by hand, but as this needs much labour, machine harvesting is increasingly used. After the cutting, the plants throw up successive crops of stems called 'ratoons', which take about the same period to mature as the original crop. The yield, however, declines slowly with each ratoon, and after two or three ratoon crops have been harvested re-planting is usually undertaken. The drain on soil nutrients is a heavy one, so that fertile soils with heavy dressings of manure and fertilizers, as well as good rainfall or irrigation, are necessary for this crop. The cane stems have a high content of cane sugar (sucrose). Before this is extracted, the leafy upper end and the leaves are removed; burning of the crop before cutting is sometimes done to lessen this work. Sugar cane plants sometimes, but not very commonly, produce their inflorescences, or 'arrows' (1A), in the field. The stems, after being harvested, must be taken rapidly to the factory with which all estates are equipped, for sugar extraction. With modern machinery this is an efficient but quite complicated process, capable of producing 'plantation white' sugar for direct consumption. In many importing countries such sugar is further refined to give a very pure product, and by-products such as 'golden syrup' are produced. In some tropical areas more primitive methods of extraction are still used, producing a crude brown sugar still containing much of the molasses; sugars of this kind are widely consumed in the tropics under such names as 'muscovado', 'gur', and 'jaggery' (India), and 'panela' (South America). More refined brown sugars are used in Europe.

Sugar cane is native to the Old World tropics, and was taken to America by Columbus. It is historically very important as the original basis of the plantation industry in the tropics, and is still largely produced on big estates though there are also small 'cane farmers' in some tropical countries; the high labour requirements of the crop were a chief cause of the development of the slave trade. Cane sugar is nowadays one of the most scientifically produced tropical products, the modern thick-stemmed cane being capable of producing more human food per acre than any other crop. Sugar cane is grown in and exported from a very large number of tropical and sub-tropical countries. Brazil and India have very large acreages. Among the chief exporters are Cuba, Hawaii, and Puerto Rico; but many smaller producers like Barbados, Mauritius, and Guyana are largely dependent on sugar exports for their prosperity. The temperate regions import very large quantities — even those countries which grow sugar beet. Rum is a by-product of sugar cane.

2 **Sugar Beet** (*Beta vulgaris subsp. cicla*) is the most important source of sugar in temperate countries which are too cold for the cultivation of sugar cane. Although the plant was known in pre-Christian times, its use as a commercial source of sugar is a modern development. About 200 years ago, a German chemist extracted 6.2% of sugar from the roots of a white variety of sugar beet. Today, using improved varieties and modern methods of extraction, yields of 15% to 20% are obtained. The sugar beet industry first became important in France and Germany early in the 19th century. Napoleon encouraged its development, as a means of boycotting cane sugar from British Colonies. In Britain, little sugar beet was grown before the 1920s, when the Government encouraged its cultivation to help arable farmers during the Depression. Large quantities of sugar beet are grown in Germany, France, Russia, Czechoslovakia, Poland, and the United States.

Sugar beet is closely related to the garden beetroot (p. 171) but has whitish, conical roots, averaging about 2 lb. in weight and about 12 to 15 inches in length. The plant is a biennial, producing a rosette of leaves and its large root in its first year, and normally not flowering until the second year — although some strains are liable to 'bolt' or flower prematurely, especially if sown too early. The crop is harvested in autumn and early winter. The leafy tops, which are cut off in the field, make excellent food for cattle and sheep, as also does the pulp which remains after the sugar has been extracted.

Sugar is extracted by shredding the roots and heating them in running water. Impurities are removed and the clear liquid is concentrated and crystallized, yielding a sugar which is indistinguishable from cane sugar. Another by-product is molasses, which is used as a stock food and for making industrial alcohol. The filter cake, left behind when the juice is purified by filtration, is used as a manure.

1

1A

2

2

1

2A

QUARTER LIFE SIZE   PLANT × 1/20
1 SUGAR CANE   1A Flowering plant
2 SUGAR BEET   2A Inflorescence

OTHER SUGAR CROPS

1 **Wild Date Palm** (*Phoenix sylvestris*). This relative of the true Date Palm (*Phoenix dactylifera*) (*see* p. 107), is planted on quite a considerable scale in India as a source of sugar. To obtain the sugar, some of the leaves covering the tender top portion of the stem are removed by a climber and a cut is made into the stem. The sap as it exudes is collected in a receptacle suspended on a bamboo. The sap is then boiled down until the residue, when cooled, consists of a thick sticky brown sugar. About 3 quarts of sap are required to produce 1 lb. of sugar. The sap is also sometimes fermented and then distilled to make the alcoholic liquid called arrack. Palms are fit to be tapped from about 8 – 10 years old, and a plantation of them may yield as much as 9 tons of this crude sugar per acre per year.

2 **Sugar Palm** (*Arenga saccharifera*). The male spadix or inflorescence is illustrated here (2). The Sugar Palm is more fully figured and described on pp. 184–5.

3 **Palmyra** or **Borassus Palm** (*Borassus flabellifer*) is here considered only as a source of sugar; its other uses are described on p. 106. It is again mainly in India that this palm is used for sugar production, though it also grows in other Asian countries and in Africa. The sap in this case is obtained by tapping the unopened spadix or inflorescence (3A and 3B); the palms are either male or female, and the female inflorescence gives a greater yield. As described above, the sap is boiled down immediately to make sugar or can be fermented to produce the moderately alcoholic drink, toddy, and then distilled to make the stronger arrack. The prohibition of alcoholic drinks in some Indian states has meant that more trees are used to produce sugar and fewer for toddy. Tapping of palms may begin about 15 years from sowing when flowers are first formed, and may continue for 30 – 40 years.

The annual tapping period is usually 4 – 5 months during the hot dry season, during which a single tree can yield a total of 50 – 80 gallons of sap.

4 **Sugar Maple** (*Acer saccharum*) is a native of north-eastern North America, where its sweet sap was used for making syrup and sugar by the Indians, long before production was commercialized. The sap of the tree begins to flow in March and continues for several weeks. The Indians tapped it by making cuts in the bark of the trees and channelling the liquid into containers through hollow reeds or curved pieces of bark. They boiled the liquid, to concentrate it, by dropping hot stones into it, and made sugar by allowing the syrup to freeze and removing the ice which formed on the surface. For local domestic production, very similar methods are still used, except that holes are drilled in the trees and fitted with spiles to carry the sap into containers. The sap is boiled in large kettles, to concentrate it to a syrup. Traditionally, the 'sugaring-off' process consists of pouring the waxy syrup on to snow and then putting it into moulds to crystallize. Commercial producers use modern evaporators, which convert the sap into syrup much more rapidly and cleanly. Maple syrup and maple sugar are mainly used in the United States and Canada, for making various kinds of confectionery, sweets, puddings, and ice-cream.

The Sugar Maple is a large tree, up to 120 feet high, bearing leaves 3 to 6 inches across, with 3 to 5 pointed lobes. Its small, greenish-yellow flowers, which have no petals, appear at the same time as the leaves.

Black Maple (*Acer nigrum*), which is closely allied to the Sugar Maple, is the most important of the other maples which are used as sources of sugar and syrup. In parts of Europe, the sweet sap of the Silver Birch is sometimes used for making birch wine.

TREES, SMALL SCALE   *SPADICES AND MAPLE LEAVES* $\times \frac{1}{8}$

1 WILD DATE PALM tree   2 SUGAR PALM male spadix
3 PALMYRA PALM tree   3A Male spadix   3B Female spadix
4 SUGAR MAPLE tree   4A Leaves

17

OIL SEEDS AND FRUITS: COCONUT PALM

Of the many plants which produce edible oils, a high proportion is grown in the tropics. Usually the 'oilseeds' are exported and the oil is extracted from them in the importing countries. The residue left after extraction, known as 'oilcake', is often a valuable livestock feed. Most of these oils are used for soap-making and other industrial purposes, as well as for food products. The edible vegetable oils should be distinguished from 'drying' oils such as linseed oil, and 'essential' oils (used in perfumery).

Coconut (*Cocos nucifera*). This palm is a common sight growing along the sea-shore of many tropical lands. Because it is tolerant of salty sandy soils, it can usefully occupy a strip of land at the top of the beach on which it would be difficult to grow any other crop. But coconut plantations can also be made in inland places, and the palm will grow on most soils that are sufficiently well-drained. It is practically confined to the tropics, but will fruit in a few warmer sub-tropical areas, as in the Bahamas. Its yield falls off when planted at any great altitude above sea-level, so plantings are confined to the tropical lowlands. Coconut palms are most abundant in the Old World, but as the earliest European explorers found them also growing in America, they had evidently become dispersed, probably by floating on the sea, at such an early date that we cannot be certain exactly where they originated. The crop is usually propagated by laying the fruits, only partly covered by soil, in a nursery; but ripe fruits will also germinate where they fall. Transplanting from the nursery into the field is done several months after sowing. The seedlings develop into handsome trees, reaching a height of up to 80 feet with a slender, often curved, trunk surmounted by the feathery crown of leaves (1). The inflorescences, and subsequently the fruits, appear amongst the leaves from about the sixth year onward; the palm may not reach full bearing until it is about 20 years old, and may go on fruiting till it is about 80 years old. The fruits are at first green (2) but turn yellow as they ripen.

With unbranched trees of such a height, the harvesting of the fruit presents a problem. Skilled climbers sometimes solve it by using a rope and taking advantage of the roughnesses left by old leaf-bases on the stem to assist them in their dangerous task. The rope passes behind the trunk of the tree, and the climber uses it either as a belt or, by making loops in its ends, as stirrups for his feet. Sometimes the harvester on the ground cuts fruits with a knife at the end of a bamboo. In some countries of south-east Asia trained monkeys, wearing a collar and long lead, are sent up the trees to throw the fruits down. In many cases, however, the fruits are left to fall naturally, but many are then too ripe. Fruits are generally produced very regularly throughout the year.

The coconut palm is a plant of many uses. The trunks are used as building timbers, and the leaves for house thatching. It is the fruit which provides the food supply. All parts of the fruit have their uses. Beneath the outer skin is a thick layer of fibrous husk (3). These fibres can be combed out and sold as coir, a material used for making ropes and coconut matting. Inside the husk is the nut, and this is what most people in temperate countries think of as a 'coconut', since it appears as an import in their countries while the whole fruit does not.

The nut has a hard shell which requires a sharp blow to break it open. At one end of the shell are three sunken 'eyes' of softer tissue (4), through one of which the young shoot and root emerge at germination. Inside the shell is a thin white fleshy layer comprising the endosperm, commonly known as the 'meat' of the coconut. The interior of the nut is hollow but partially filled with a watery liquid called 'coconut milk' which is most abundant in unripe nuts, and is gradually absorbed as ripening proceeds. It is a refreshing and nutritious drink, containing sugars and a little oil. The harvested fruits are first split open and the husk removed, typically on a sharpened iron stake stuck into the ground. The nuts are then broken open with a cutlass. The meat is sometimes eaten directly as a food, especially in the Pacific islands where the number of available foodstuffs is small. The meat may be used to prepare desiccated coconut. But by far the most important economic product of the coconut palm is obtained by drying the white meat into copra. The meat is detached in two halves from the nut and dried either in the sun or in a kiln which may be heated by burning the coconut shells. The dried copra is the source of coconut oil. Some extraction is done in the coconut-growing countries, but there is a very large export trade in copra to countries which extract the oil themselves. Good copra contains 60–65 per cent of oil. The primary use of coconut oil has always been for soap-making, a field limited in recent times by the competition of detergents. Its main uses for food purposes are as a cooking oil and in margarine manufacture. The residue left after the extraction of oil from copra is a valuable oilcake for feeding livestock. The country which exports by far the most copra is the Philippines; among the more important of the lesser exporters are Indonesia, India and New Guinea. Copra exports are particularly important to many Pacific islands which, while producing only small quantities, are economically very dependent on them.

In many tropical countries the sap of the coconut palm is tapped by cutting the end of the unopened spathe surrounding the young inflorescence. This sap has a high sugar content and can be evaporated down to make crude sugar, or fermented to produce a popular drink called 'toddy'. Toddy can be distilled to produce a spirit called 'arrack'.

TREE SMALL SCALE   DETAILS × ⅛   OPENED NUT × ⅔

1 COCONUT PALM

2 Immature fruits   3 Opened ripe fruit and seed
4 Opened nut   5 Young plant

19

OIL PALM

Oil Palm (*Elaeis guineensis*). This is one of the world's most important sources of edible and soap-making oil. As it yields more per acre per year than can be obtained from any other vegetable oil or from animal fat, its importance is not likely to decline. The palm originated in western tropical Africa, where it grows wild in large numbers often constituting 'palm groves', especially where the virgin forest has been felled. A large proportion of the commercial supplies of this oil arises from fruits collected from such self-sown trees by the country people of West Africa; but plantations have also been established, for example in the Congo and the Ivory Coast, and in Malaysia and Indonesia. The palm will thrive only where rainfall is fairly high, but is a valuable crop for making an economic use of rather poor soils.

The oil palm is planted from seeds, but as they do not germinate easily they are often subjected to heat treatment before sowing. The seedlings are raised in a nursery and planted out into the field at about 12 – 18 months old. Bearing begins at about 5 years, reaches a maximum at about 10 years, and the trees may remain in the ground up to 50 years. Male and female flowers are borne in separate inflorescences on the same tree, but do not mature at the same time, so that cross-fertilisation by pollen from another tree is the rule. Artificial pollination is sometimes carried out on plantations to improve the fruit set. The female inflorescence develops a large fruit bunch containing up to 200 fruits, which when ripe are most commonly of an orange colour but in some varieties black or of other shades. A tree may produce 2 – 6 bunches of fruit a year. Harvesting the mature trees, which may be 30 – 50 feet high, is usually done by climbers who use a rope round the trunk of the palm passing behind their waist or forming stirrups for their feet to assist them.

In order to understand the processing of oil palm fruits, it is necessary to know something about their composition. Beneath the outer skin of the fruit is a layer of fibrous pulp, called the mesocarp or pericarp (4A), which is rich in oil; this is the palm oil of commerce. Inside this is the seed (commonly called the 'nut'), consisting of an outer hard black shell which is useless except for fuel, and an inner kernel which is also rich in the more valuable 'palm kernel oil' which is suitable for making margarine. The simple method of processing used by many small African farmers is to leave the bunches to ferment for a few days and then to strip off the fruits and boil and pound them. The palm oil from the mesocarp is obtained by skimming the surface of the water or by squeezing and washing the fibre. After further settling and heating to remove water the palm oil is ready for use or sale.

A simple improvement on this process is to use a hand-operated oil press. The nuts in either case still remain, and their shells can be broken between two stones or in a hand-operated cracking machine. This exposes the kernels, which are sold on a large scale, for export and extraction of their oil in the importing country. On plantations very different processes are used because they have a factory; and in some parts of West Africa factories have also been erected to which small producers can bring their palm fruits. By factory processes, both palm oil and palm kernel oil can be extracted on the spot, and a larger percentage of oil is recovered from the fruits than by hand processing.

The amount and proportion of the two oils produced also depends on the variety of oil palm used. Most of the wild oil palms of West Africa are of the variety *dura* which has a very thick shell and a rather thin mesocarp, so that its yield of palm oil is low. African plantations usually grow the better variety *tenera* which has a much thicker mesocarp and higher yield of palm oil, though its kernels tend to be small. The south-eastern Asian countries have their own good variety 'Deli' which is generally grown in Indonesia and Malaysia, and also a 'dumpy' variety which is easier to harvest because the palms do not grow so tall.

Palm oil before refining is of a yellow or orange colour due to the pigment carotene from which the human body can form vitamin A. This gives it a special value in the diet. Although liquid when first extracted, it sets as a very soft solid at normal temperatures. Perhaps half the quantity produced in an African country such as Nigeria is used locally as food in all kinds of cookery, and the rest exported; in Asia it is hardly eaten at all. When exported, palm oil is used largely for soap-making and industrial purposes; but it can if suitably treated be used to make margarine. Palm kernel oil is white or pale yellow and more largely used for margarine making. Its extraction from the kernels leaves a residue of oilcake which is a good livestock feed. As with many other palms (*see* p. 16), a 'wine' may be prepared from the oil palm by tapping and fermenting the sugary sap. It is appreciated in West Africa and has a useful content of vitamins of the B group.

The four countries producing most palm oil and palm kernels or palm kernel oil are Nigeria (in the southern high rainfall belt, and especially in the south-east), Malaysia, the Congo, and Indonesia. Somewhat behind these major producers are Ghana, Liberia, Sierra Leone, and a number of other African countries. Several countries of tropical America are developing production although it is still small.

2

2A

3A

3

4A

4A

4

MALE SPADIX × ¼   *FRUITING SPADIX* × ½   *FLOWER AND FRUIT DETAILS* × 1

1 OIL PALM (small scale)

2 Male spadix   2A Detail of male flowers

3 Branch of female flowers   3A Detail of female flower

4 Fruiting spadix   4A Details of fruits and nut

OIL SEEDS AND FRUITS: OLIVE, SESAME, PEANUT

1 **Olive** (*Olea europaea*). The fruit of this plant has been valued since ancient times, not only as food but also as the source of an edible oil. This oil was also used in lamps — in some religions it is prescribed for burning in sanctuary lamps — and for anointing, and for many other purposes including medicinal and cosmetic uses. As an edible oil, it is used chiefly for cooking, as a salad oil, and for canning sardines. The olive originated in the Mediterranean region (there are many references to it in the Bible and in classical writings) and many countries in that region, particularly Spain, France and Italy, are important producers. The tree is also grown in California, South Australia, China, and other sub-tropical and warm-temperate areas, usually in fairly arid regions, or well-drained soil.

Olive oil is obtained by pressing the ripe fruits, by methods which vary in different countries. The finest grades, used for culinary and medicinal purposes, are a clear, golden yellow and almost odourless. Inferior grades are greenish-yellow and are used for making soap and as lubricants.

The olive is a small, slow-growing tree which often lives to a great age, its grey trunk becoming very gnarled and picturesque. It begins fruiting when it is several years old, and continues for very many years, bearing the fruit only on the wood of the previous year. The lanceolate leaves, 1 to 3 inches long, are green above and silvery-scaly beneath. The fruit is a drupe, green at first and dark blue or purplish when ripe, containing a single hard seed. The fruits are picked when ripe, or when fully grown but still green, and are usually pickled in brine. They are much used in *hors d'oeuvres* and in various dishes, especially in the regional cookery of Mediterranean countries. Stuffed olives are green, pickled olives, with the seed removed and replaced by a piece of pimento or sweet red pepper.

2 **Sesame** (*Sesamum indicum*) is also known in English by more local names, such as 'sim-sim' in East Africa and 'benniseed' in West Africa. It is a crop of African origin, now also common in tropical and sub-tropical Asia where it extends to the Mediterranean area. It is an annual, propagated by seed and taking 3 – 5 months to mature. The plants grow up to 6 feet high; the flowers in different varieties are white, pink, or mauve. In harvesting, the whole plants are cut and stacked in an upright position, often against a rack; as they dry, the seed capsules split open at the apex and the plants can then be turned upside down and the seeds shaken out on to a cloth. The seeds are small (2A) and white in the varieties most commonly grown. They contain from 45 to 55 per cent of oil, which is made use of in various ways in human diets. Many tropical families stew the seed whole; whole seeds are also used in sweetmeats in Asia, and in some countries to decorate bread and cakes. Where the oil is extracted commercially, it is used as a cooking and salad oil and also for making margarine. Compared to other oil-seeds, sesame is a rather low-yielding crop and its acreage in recent years has remained stagnant. The largest producing countries are, in order of importance: India, China, Burma, and the Sudan. Nigeria also has an export of sesame. The sesame cake left after extraction of oil is an important protein-rich stockfeed.

3 **Groundnut** or **Peanut** (*Arachis hypogaea*). The plant, whose seeds are sometimes known as 'monkey-nuts', is a member of the family Leguminosae. Of South American origin, it is now a most important crop all over the tropics and sub-tropics and is widespread in the United States as far north as Virginia. It is an annual, propagated by seed, with a crop period of 4 – 5 months. In some varieties the plants are erect (3B), and in a good crop may reach knee height; in others the stems are prostrate, spreading far over the surface of the ground and rooting at the nodes. The flowers are yellow, and usually self-pollinated, some of them actually being borne below ground level. After pollination, the stalk bearing the flower elongates and forces the young pod down into the soil where it completes its development. Groundnuts therefore have to be dug out of the soil at harvest, like a root crop. In the erect varieties, the pods are well grouped round the central stem (3B), but in the runner varieties they are much more scattered and laborious to harvest. The crop does best on a light soil, which also makes harvesting easier; the same object is sometimes attained by growing it on ridges. Most of the world's groundnuts are dug by hand, but in the United States machines are used to lift the crop.

The groundnut pod, with its wrinkled surface network, contains usually 2 nuts or kernels in the Spanish or Virginia types, 3 – 4 in the Valencia types. The nuts can be shelled out of the pod by hand or in hand-operated or powered machines. They are extremely nutritious because of their high content of both protein (about 30 per cent) and oil (40 – 50 per cent); they are also rich in vitamins B and E. In tropical households the whole nuts are used in cookery; and whole nuts, of selected large dessert types, with the reddish skin removed, are also eaten raw or roasted in temperate countries. Peanut butter is made by removing the skin and germ and grinding the roasted nuts. There is, too, a large export trade in groundnuts for the commercial extraction of oil. Besides being used as a cooking and salad oil, groundnut oil is one of the most important fats used for making margarine. It is also used for packing fish, for example, tinned sardines. The residue left after oil extraction is one of the most used oilcakes for animal feeding. The country producing most groundnuts is India, which has little to spare for export, and is followed by China. Nigeria is the largest exporter of groundnuts, which are grown in the drier northern part of the country. Groundnut exports are very important in the economy of certain smaller countries, such as Senegal and Gambia. Groundnuts are one of the prime bases of the oilseed import trade which is a speciality of certain ports in Europe such as Marseilles, Liverpool, and Hamburg, and the oil extracting factories are often concentrated near these ports.

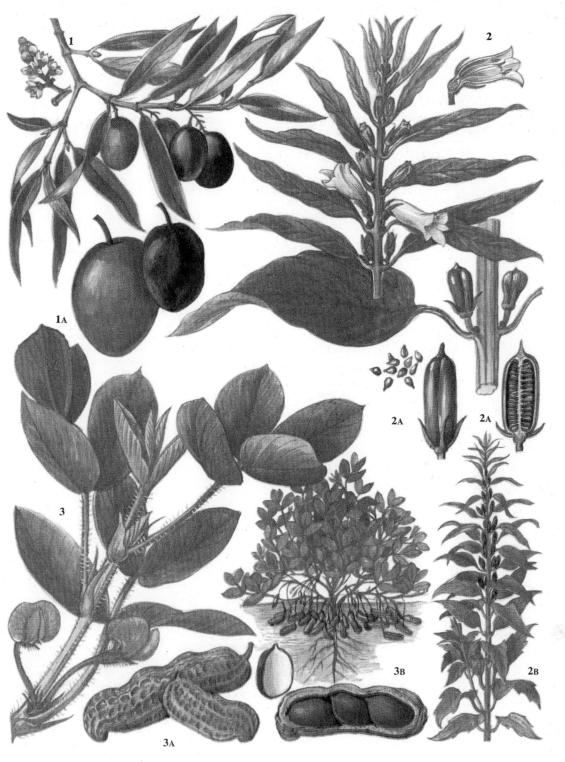

LEAFY SHOOTS × ⅔   *FRUITS LIFE SIZE*   *PLANTS* × ⅛

1 OLIVE flowers and fruiting branch   1ᴀ Fruits

2 SESAME flowering stem   2ᴀ Seed pods and seeds   2ʙ Part of plant

3 PEANUT (GROUNDNUT) flowering shoot   3ᴀ Pods and nuts   3ʙ Plant

OTHER OIL-PRODUCING PLANTS

1 **Soya** or **Soybean** (*Glycine max*). The seeds yield an edible oil which, when refined, is used for cooking, in salads, and for making margarine. This oil is also important in industry, being used in the manufacture of soap, paints, plastics, and many other products. Although for convenience considered here with other oil-producing plants, soya would have at least an equal right to appear on the pages concerned with the important beans. It is used fresh, fermented, or dried, as a vegetable. It is one of several beans (*see* p. 38) used in Chinese cookery as bean-sprouts (young seedlings grown in the dark are picked about a week after germination before the first true leaves have expanded). It is the source of a flour which is used mixed with wheat flour in bakery, and also as an ingredient of ice-cream and many other food products. The flour has a high protein and a low carbohydrate content. Soybean 'milk', extracted from the seeds, is used in cooking in China and Japan and is recommended as an invalid food. Soy sauce, a dark brown liquid sauce, is provided in every Chinese restaurant. Both the soya plant, and the residual oilseed after oil extraction, are fed to livestock.

Soya is believed to be native to south western Asia, where it has been grown since ancient times. It is the most important food-legume in China, Manchuria, Korea, Japan, and Malaysia. Large quantities are grown in the U.S.A. and in other parts of the world with similar climates. It is not an economic crop in the British Isles, as it is susceptible to frost and our summers are rarely warm enough to produce high yields. Over a thousand varieties are known, so the description of the plant must be generalized.

Soya is an annual herb belonging to the Pea family (Leguminosae). It is an erect, bushy plant, 1½ to 6 feet high, bearing rough, brownish hairs on its stems, leaves, and pods (1B). Its leaves have three broad leaflets, 2 to 6 inches long. The white or purple flowers are papilionaceous, with a rather broad standard and smaller wings and keel. The hairy pods, up to 2 or 3 inches long and ½ inch wide, are constricted between the 2 to 4 seeds (1A), which may be green, brown, yellowish, or black.

2 **Sunflower** (*Helianthus annuus*). An edible oil is extracted from the 'seed' and used for cooking and in salads, margarine, and other foodstuffs. It is also used in varnishes and soaps. The residual oilcake is a valuable cattle-food and the stems and leaves are also used for fodder, either fresh or as silage. The flowers are attractive to bees and therefore are an indirect source of food, in the form of honey. Sunflower seeds are eaten raw in some countries, and are also fed to poultry and many cage-birds. The crop is grown chiefly in temperate countries including eastern Europe, Argentina, and Canada, but is of little importance in the British Isles. It is sometimes grown in the tropics. The Annual Sunflower belongs to the Daisy family (Compositae). It probably originated in western North America or Mexico. It is a stout-stemmed, often unbranched, erect herb, 3 to 9 feet or more in height, coarsely hairy on its stem and on its large, spirally arranged, long-stalked, ovate leaves. The lower leaves are heart-shaped, the smaller, upper leaves are square-based or tapering at the base. The large, terminal flower-heads, sometimes more than 12 inches across on cultivated plants, are composed of bright yellow asexual ray florets and brownish, tubular, bisexual disc florets. There are garden varieties with flowers of various shades, from pale yellow to chestnut brown. The 'seeds' (2A), which botanically are really dry, indehiscent fruits (achenes) are also variable in colour — white, brown, black, or often dark with white stripes. The Annual Sunflower was introduced into Europe in the middle of the 16th century. It is a hardy annual and able to grow in most soils. The seed-heads are cut when ripe and either hung up or left on the ground to dry. The latter method is easier but is liable to result in an inferior quality of oil.

3 **Rape** (*Brassica napus*). The plant is of uncertain origin, cultivated since ancient times and closely related to the garden swede. Rape, also known as 'Cole' or 'Coleseed', is the source of an edible oil, pressed from the seeds of annual varieties and used for culinary purposes in Europe and in India. One of its uses is for oiling loaves of bread before baking. Rape is widely grown for feeding to cattle. It is also used for winter and spring salads and for spring greens. The seed-cake is used for cattle food.

Rape is an annual or biennial herb, belonging to the Wallflower family (Cruciferae). It has erect, branching stems up to about 3 feet high, bearing deeply lobed, grass-green, bristly lower leaves, and less deeply lobed, glaucous, bluish-green, more or less glabrous upper leaves, the uppermost being unlobed and clasping the stem with their deeply cordate base. The inflorescence is racemose, with numerous yellow flowers, each with the 4 sepals, 4 petals, and 6 stamens characteristic of the family. The inflorescence does not lengthen so that the opened flowers rise above the buds. The fruit (3B) is a brownish siliqua about 2 to 4 inches long, with a long, tapering beak, and containing blackish or reddish-brown seeds (3A) attached to a thin, whitish, false septum.

Other Oil Plants which deserve mention include cotton, from whose seeds an edible oil is extracted although fibre is the crop's main product; and some wild plants whose fruits are collected for oil extraction when prices are sufficiently attractive, such as the palm *Orbignya* in Brazil and the shea butter-nut (*Butyrospermum*) in tropical Africa. Two minor annual oil-seed crops, grown mainly in India but also in some other countries, are safflower (*Carthamus tinctorius*) and niger seed (*Guizotia abyssinica*).

PLANTS × ⅛   SEEDS LIFE SIZE   DETAILS × ⅔

1 SOYA BEAN plant   1A Seeds   1B Details of flower, leaves and pods
2 SUNFLOWER plant   2A Seeds
3 RAPE plant   3A Seeds   3B Detail of ripe fruits

25

NUT TREES OF TEMPERATE CLIMATES

The term 'nut' is used here in its popular sense to describe any seed or fruit consisting of an edible and usually rather hard and oily kernel within a hard or brittle shell.

1 Hazel or **Cob** (*Corylus avellana*). The wild hazel of Asia Minor and Europe is a bush or small tree whose nuts have been eaten by man since the earliest times. They were collected for food by Mesolithic peoples, and there are references to them in Theophrastus and Pliny. Many of the existing cultivated varieties originated during the last century, when there was considerable interest in selection and hybridisation. Although hazels grow in most parts of Britain, nut production on a commercial scale is restricted to Kent, where cobs and filberts are grown mainly on the Lower Greensand formation. The trees start bearing fruit when about 6 years old, reaching peak production in about 15 years and continuing to crop profitably for 50 years or more. Harvesting begins in September. Kent cannot satisfy the demand for nuts for dessert and for confectionery. Consequently, thousands of tons of cob nuts and the closely related filberts and Turkish hazels are imported annually. Major producing countries include Spain, Italy, France, Turkey and Armenia.

The hazel tree grows up to 20 feet high, but is often restricted by pruning. The male catkins, 1 to 3 inches long (1A) are conspicuous by their bright yellow anthers, from January to April. The female flowers are less noticeable, occurring in short, bud-like spikes, with their crimson styles exposed. The nuts, in clusters of 1 to 4, are globose or ovoid $\frac{1}{2}$ to $\frac{3}{4}$ inch long, with a hard brown shell, partially enclosed by a deeply lobed involucre or husk (1B).

2 Filbert (*Corylus maxima*). This species, a native of south-east Europe, is believed to be the parent of most of the cultivated filberts and sometimes occurs naturalized in countries where it was originally planted. The distinctions between cobs and filberts have been obscured in the course of breeding and selection. The well-known 'Kentish Cob' is really a filbert ('Lambert's Filbert').

The filbert is more robust than the hazel, growing to a height of 20 feet or more. Its nuts are usually oblong ovoid in shape, but the most reliable distinction between the two species is to be seen in the husk or involucre, which in the filbert extends well beyond the nut and is constricted at the apex.

3 Sweet Chestnut (*Castanea sativa*). This native of southern Europe is extensively planted elsewhere for its nuts and timber. The tree was probably introduced to Britain by the Romans. The name 'Spanish Chestnut' is no doubt derived from the large quantities of nuts imported from Spain. In the warmer parts of southern Europe, the chestnut has been grown for many centuries, the nuts being ground into flour and used in soups, fritters, porridges, stuffings, and stews, as well as being eaten whole, either boiled or roasted, and preserved in sugar or syrup, as in the famous French delicacy, marron-glacé. In some areas, the nuts are commonly fed to livestock, chestnut-fed pork being highly esteemed. In Britain, nut production is of little importance, partly because the summers are too unreliable, partly because the trees are usually seedlings, not selected cultivars. In southern Europe, where nut production is taken seriously, cultivars are budded or grafted, to produce superior nuts.

The sweet chestnut is a large tree, up to 115 feet high, with a broad crown. The leaves are 4 to 10 inches long, oblong lanceolate, and coarsely toothed. The male and female flowers are borne in separate inflorescences, in July. The male catkins, 4 to 8 inches long, are conspicuous because of their yellow anthers (3A). The female flowers are usually borne in threes, each with 7 to 9 red styles. The glossy, brown nuts, $\frac{3}{4}$ to 2 inches wide, are enclosed in a green cupule densely covered with long, branched spines.

4 Almond (*Prunus dulcis*). This small tree probably originated in the Near East, but is now naturalized in many areas of southern Europe and western Asia, where it is cultivated. It has been introduced successfully into other parts of the world where the climate is suitable, including California, South Australia, and South Africa. There are two varieties of economic importance. *Prunus dulcis* var. *amara*, the bitter almond, is the chief source of almond oil which is used for flavouring and in emollient preparations for the skin. The kernels of the nuts and the crude oil, before refining, contain prussic acid (HCN), but the bitterness of the nut should deter anyone from eating enough to be poisoned. *Prunus dulcis* var. *dulcis*, the sweet almond, is the variety grown for its edible nuts. It includes numerous cultivars, differing not only in the characteristics of their nuts but also in flower colour, habit, size and shape of leaf, and ecological adaptations. The Jordan almond, grown in south-east Spain, is unprofitable elsewhere, being generally a shy bearer. Important producing countries include Italy, Spain, France, and California. Apart from being popular dessert nuts, almonds are much used by bakers and confectioners. Of all the nuts, almonds have the largest share of world trade.

The almond resembles the peach, *Prunus persica*, in its pink flowers, in fascicles of 1 to 3, appearing before the leaves; but its flowers are often slightly larger, up to 2 inches in diameter. The leaves of both species are folded in bud, but those of the almond are broadest rather below the middle and more minutely toothed (4A), whereas the leaves of the peach are broadest about or above the middle (*see* p. 73). The fruit of the almond is usually smaller than most peaches, green and leathery when ripe and splitting open to reveal the 'nut' or stone. Almonds, like several other rosaceous fruit trees, are often self-sterile, requiring pollination by another cultivar before nuts can be formed.

NUTS LIFE SIZE   *BRANCHES* × ¼

1 WILD HAZEL NUT   1A Flowering and fruiting branches   1B Cultivated COBNUTS
2 FILBERT NUTS   2A Fruiting branch
3 SWEET CHESTNUTS   3A Flowering and fruiting branches
4 ALMOND NUT   4A Flowering and fruiting branches

NUT TREES OF TEMPERATE CLIMATES

1 **Walnut** (*Juglans regia*). This beautiful tree is native from south-eastern Europe to West and Central Asia and China. It has been cultivated in the British Isles for so long that the date of its introduction is doubtful, although it may have been during the 15th century. It is valued as a timber tree as well as for its nuts which are excellent for dessert and also in baking and confectionery. Walnut oil, extracted from the nuts, was formerly of considerable importance as an edible oil and has been used for centuries in the preparation of artist's paints. The young green fruits, gathered before the 'nuts' harden, are eaten pickled in vinegar. Trees grown from seed cannot be relied upon to yield good quality nuts, which are usually obtained from grafted or budded trees. France is the leading producer of walnuts but Italy, Roumania, California, and China are also important producers and the tree thrives in many other countries.

The Walnut is typically a tall deciduous tree, up to about 100 feet in height, with grey bark, smooth on young trees and fissuring with age. Its leaves are alternate, pinnate, with 5 to 11 (rarely 13) obovate or elliptic leaflets, 3 to 6 inches long, the terminal leaflet largest. The male flowers, with a small, lobed perianth and 3 to 40 stamens, are borne in long, pendulous catkins (1B). The female flowers are solitary or few in number. The fruit is a green drupe (1), containing a wrinkled stone (walnut).

2 **Black Walnut** (*Juglans nigra*). This is a native of North America, introduced into Britain during the seventeenth century. It has nuts which are often larger than the average European walnut, but the shell is usually too thick and hard to be cracked with ordinary nutcrackers — in America, special nutcrackers are made for the purpose. The kernel is of good quality, although rather strong in flavour, and is used in confectionery, ice-cream, etc. in the United States. Several named cultivars have been selected for their thinner or softer shells.

The Black Walnut grows up to 150 feet in height. It has dark brown, furrowed bark, and pubescent branches. Its dark green leaves (2A), 1 to 2 feet long, have 11 to 23 serrate leaflets, each 2 to 5 inches long.

3 **Butternut** (*Juglans cinerea*). Also known as 'White Walnut', this is a rather smaller North American tree, bearing fruits about 2 inches long and 1 inch across,

in groups of 2 to 5. The shell, though hard, is generally not difficult to crack, and the kernel has a rich and pleasant flavour. Some cultivars have been selected for their improved shelling qualities.

The Butternut grows occasionally up to 100 feet high. It has grey bark and pubescent branches. The pinnate leaves (3A) have 7 to 17 irregularly serrate leaflets, glandular hairy, 2 to 5 inches long. The ellipsoid fruit is also sticky with glandular hairs.

Some other species of *Juglans* are of limited local importance as sources of edible nuts — for example, the hardy, Japanese Walnut (*Juglans sieboldiana*).

4 **Pistachio** (*Pistacia vera*). The small tree, native to the Near-East and Central Asia, has long been cultivated in the Mediterranean region and more recently in the southern United States. The green kernels are highly prized, as much for their ornamental colour as for their pleasant, mild flavour, and they are much more expensive than most nuts in commerce. In producing countries they are commonly eaten salted, like peanuts, as well as being used in confectionery. Elsewhere, they are mainly used for decorating and flavouring ice-cream, nougat, cakes, trifles, etc.

The Pistachio is a deciduous tree, about 20 to 30 feet high. Its leaves have 3 to 7 ovate leaflets, 2 to 4 inches long. Male and female flowers are borne on different trees, in axillary racemes. They are small and have no petals. The fruit is an ovoid drupe.

5 **Pecan** (*Carya illinoensis*). This North American nut has been long appreciated in its own country, where large quantities are eaten as dessert nuts, plain or salted; in ice-cream, cakes, nut bread, candies, and other confectionery; and in vegetarian croquettes and sandwiches. Nowadays, pecans can be bought in many British shops, but they have not yet succeeded in rivalling the popularity of the old-established favourites, such as Brazil nuts, almonds, walnuts, and hazels. *Carya* belongs to the same family as *Juglans* and pecans have a mild, sweet walnut-like flavour.

The Pecan is a large tree up to 170 feet high, with grey, furrowed bark. Its leaves are pinnate, with 7 to 17 leaflets, 4 to 7 inches long. The male flowers are borne in 3-branched, pendulous catkins, the female in 2 to 10 flowered spikes. The fruit is a drupe, $1\frac{1}{2}$ to $3\frac{1}{2}$ inches long, which at maturity splits into 4 valves to reveal the smooth, brown kernel.

1 EUROPEAN WALNUT   1A Fruiting shoot   1B Female flowers and male catkin
2 BLACK WALNUT   2A Leaf   3 BUTTERNUT   3A Leaf
4 PISTACHIO   4A Fruiting branch
5 PECAN   5A Flowering branch   5B Fruiting branch

NUT TREES OF WARMER CLIMATES

1 Brazil Nuts (*Bertholletia excelsa*) are the seeds of a tall forest tree of South America. The tree bears large, hard, woody fruits, each weighing 2 to 4 lbs. The fruit has an aperture at the end which is closed by a woody 'plug', and has to be cut or broken open to extract the 12 – 24 nuts which may be found inside (1A). A similar fruit is found in the related sapucaya nut (*Lecythis zabucajo*), another South American tree whose nuts are considered by many to be superior to the Brazil nut, though not so well known in commerce. The individual Brazil nuts have to be cracked to obtain the large white-fleshed kernel which is the part eaten. With 66 per cent fat (one of the highest figures for all nuts) and 14 per cent protein, the kernels are a rich and highly nutritious food. Commercial supplies of Brazil nuts are derived entirely from wild trees, chiefly in Brazil and Venezuela. Besides being used for local consumption, the export trade takes over 50,000 tons of nuts a year and is directed chiefly to the United States and Europe. The nuts, eaten raw, are a popular delicacy, appreciated in Britain especially at Christmas time.

2 Cashew Nuts (*Anacardium occidentale*) are the product of a medium-sized tropical tree, native to America but now also extensively cultivated in India and eastern Africa. It has the virtue of growing in rather drier areas than most trees which give economic products. The fruit (2A) is a curious one, consisting of a large fleshy 'apple', below which hangs the true fruit, a single nut whose kernel is the edible part. The demand for cashew nuts has risen enormously during this century. They contain 45 per cent fat and 20 per cent protein and are used as dessert nuts or served with cocktails. After picking, the nuts have to be roasted and then shelled, a tedious process by hand labour; machines have been invented to do this task, but are not yet perfected. India has a large number of cashew nut 'factories', and some East African nuts are sent to India to be processed there before export. The main exporting countries are India, Mozambique, and Tanzania; the largest importer is the United States, but cashew nuts are also well known in Europe. In some countries a fermented liquor, 'kajú', is made from the cashew 'apples'. The shells of the nuts contain an oil which is irritant to the fingers; it is extracted and used for waterproofing and preservative puposes, and also, when polymerised, in various industrial manufactures.

3 Pine Kernels (*Pinus pinea*) are the seeds of the Stone Pine, a native of the Mediterranean region. They are eaten, like peanuts, either raw, or roasted and salted. In Italy, their traditional uses are in soups and ragoûts. They are also used in chocolates, marzipan and other confectioneries, and in vegetarian foods. They have a softer texture than the true dessert nuts. The Stone Pine (3B) is a picturesque, umbrella-shaped tree, up to 80 feet high. It bears glossy, brown, roundish cones (3A), about 4 to 6 inches across, which have to be exposed to the heat of the sun to make their scales open so that the seeds (3) can be extracted. Each seed is encased in a hard shell, which is removed by mechanical crushing to release the ivory or yellowish-white kernel.

The seeds of several other Pines are eaten, in various parts of the world: for example, the Arolla Pine (*Pinus cembra*), in Switzerland; Pinus sibirica, in Russia; the Mexican Nut Pine (*Pinus cembroides*), in Mexico and south-western America; and Gerard's Pine (*Pinus gerardiana*) in the Himalayan region. All these pines can be grown in favourable parts of the British Isles, but their edible kernels are rarely collected here.

The seeds of the Chile Pine or Monkey-Puzzle (*Araucaria araucana*) are eaten in its native Chile, as are those of the Parana Pine (*Araucaria angustifolia*) in Brazil, and the Bunya-Bunya Pine (*Araucaria bidwillii*), in Queensland.

4 Queensland Nuts (*Macadamia ternifolia*) are native to north-eastern Australia, where their flavour as dessert nuts was first appreciated; but commercial cultivation has developed in Hawaii. The tree grows to a height of about 45 feet and bears fruits containing a single nut, whose shell is rather hard to break. Factory processing in Hawaii consists in cracking the shell and grading, roasting, and salting the nuts, which have a fat content of over 70 per cent. Most of the crop is sold in the United States as dessert nuts, but a small proportion is used by bakers in confectionery.

5 Moreton Bay Chestnut (*Castanospermum australe*) is only of local importance in its native Australia, where its seeds are collected by the aborigines, who soak them in water before drying and roasting them. If eaten fresh, the seeds may be harmful.

The Moreton Bay Chestnut is a large evergreen tree, belonging to the Pea family (Leguminosae). Its leaves are compound, with 11 to 15 leaflets. Its yellow, orange or reddish flowers are succeeded by pods up to 9 inches long, each containing several large, brown seeds (5A).

NUTS AND SEEDS LIFE SIZE   FLOWERS AND FRUITS × ¼

1 BRAZIL NUT and kernel   1A Fruit   2 CASHEW NUT and kernel   2A Cashew apples
3 PINE KERNELS   3A Pine cone   3B Stone Pine tree (small scale)
4 QUEENSLAND NUT   4A Fruit and flowers
5 MORETON BAY CHESTNUT   5A Flowering branch and pods

31

EXOTIC WATER-PLANTS USED AS FOOD

1 **Lotus** (*Nelumbium nuciferum*) is the sacred lotus of India and China. It is a water-plant whose flowers and bell-shaped leaves are carried well above the water surface on long stalks, and it grows wild throughout Asia south of a line from the Caspian Sea to Manchuria. Sculptural remains show that it was also grown in ancient Egypt, but it is not now found there in the wild state. The parts which are most frequently collected for use as food are the rhizomes and seeds. These are used most commonly in times of food scarcity, but rhizomes are also sold in shops in some parts of south-east Asia. The rhizomes, which should be used when fairly young, are roasted or steamed or sometimes pickled, and have a taste something like artichokes; in China a kind of arrowroot is sometimes prepared from them. The seeds are usually boiled or roasted after removing the embryo which has a bitter taste; they can also be eaten raw. Other parts of the plant which are sometimes eaten are the fruit (1A) after removal of the seeds, the flower stem, and also the leaves which can be eaten raw as a salad.

Other plants to which the name 'lotus' is sometimes applied are the species of *Nymphaea*, or water-lilies. These belong to the same botanical family as *Nelumbium*, the Nymphaeaceae, but can be distinguished from it by the fact that their leaves float on the surface of the water. Different species of the genus grow wild almost all over the world, and use of their rhizomes as food in times of famine has probably also been world-wide. The most important species from the food point of view is *Nymphaea lotus* with pink to crimson flowers, which is native to Africa where it is most used as a food source, but has also spread to many other regions. The rhizomes can be roasted, or dried and ground into meal. The seeds are also eaten after being roasted, made into sauce, or ground to meal; and the receptacle is occasionally eaten. The rhizomes and seeds of *Nymphaea stellata*, a smaller species with blue flowers, are similarly used as a famine food in India.

2 **Water Chestnut** or **Caltrops** (*Trapa natans*) has an edible seed which has been used for food since Neolithic times. It is still eaten, in parts of Central Europe and Asia, raw or roasted, like an ordinary chestnut, or boiled. It has a floury texture and an agreeable flavour.

The Water Chestnut belongs to the Willow-herb family (Onagraceae). It is an attractive aquatic plant, with two kinds of leaves; finely divided, feathery, submerged leaves and broadly diamond-shaped floating leaves which are usually mottled or variegated. The small white flowers are succeeded by a hard, dark-grey fruit, 1 to 2 inches across, bearing two opposite pairs of large, horn-like projections. (Caltrops, an ancient weapon of war, likewise had four spine-like projections.)

Related species are used for food in other parts of the world. Ling (*Trapa bicornis*) is grown in China, Korea and Japan. Its seeds are eaten boiled, in various regional dishes, or preserved in honey and sugar: their use as food is often associated with Chinese festivals and they are also used for making flour. As the botanical name suggests, the fruit of this species bears only two 'horns'. Singhara Nut (*Trapa bispinosa*) is a native of tropical Asia. It is extensively grown in the lakes and pools in Kashmir. The seed is eaten raw or cooked, either whole or as a kind of porridge. The fruit may have either one or two 'horns'.

3 **Chinese Water Chestnut** or **Pi-tsi**, is the tuber of a sedge, *Eleocharis tuberosa*, not a species of *Trapa*, as might be supposed from its name. The plant belongs to the Sedge family (Cyperaceae). It spreads by means of horizontal rhizomes, in shallow water at the edges of lakes and in marshes. Its tubular, rush-like leaves and stems grow erect, in clumps, arising from a squat, dark-brown, basal tuber or corm, which is the part used for food in the East Indies, China, and Japan, where the plant is cultivated. Canned Chinese water chestnuts are imported into Britain from Hong Kong. They are usually sliced for use, in Chinese soups and other dishes. As with bamboo shoots, they are appreciated as much for their crisp texture as for their flavour.

PLANTS × ⅛   DETAILS LIFE SIZE

1 LOTUS   1A Opened seed head and seed
2 WATER CHESTNUT   2A Flower   2B Fruits
3 CHINESE WATER CHESTNUT   3A Flowers   3B Base of plant   3C Corms

EXOTIC LEGUMES

1 **Pigeon Pea** (*Cajanus cajan*). This plant is rather unusual among tropical legumes used for human food in that it is a shrub and a perennial plant, although a short-lived one. Probably native to Africa, it had also reached tropical Asia in prehistoric times. The crop is planted from seed, sometimes in mixed cultivation with other crops. It develops a deep rooting system and has good drought-resistance, making it suitable for relatively dry areas. African types grow 5 – 6 feet high, Indian ones somewhat taller. From the yellow flowers there develop pods usually containing 3 – 4 seeds which are of varying reddish shades with a prominent white hilum (1B). The crop is usually harvested for mature seeds, which may be ready from 5 – 12 months after planting, according to variety; but immature seeds are sometimes picked for immediate cooking. The yield falls off after the first year, and although the plants may live for up to 5 years, they are rarely kept for as long as that and are very often simply grown as an annual. Pigeon peas are very widely grown in India, where they are also known as 'red gram' and are the second most important pulse (after chick peas); the split seeds are cooked to provide 'dhal', the typical pulse dish of India. The crop is quite a popular one in the West Indies, where pigeon peas provide a useful part of the protein supply in the diet of the poorer people, and a canning industry has been developed. The crop is grown on a smaller scale in many other parts of the tropics, but is not suited to the wettest areas. Because of its long growing period and sensitivity to frost, it is not cultivated outside the tropics. Both the pods and the foliage of the pigeon pea are used in some countries as animal fodder.

2 **Bambarra Groundnut** (*Voandzeia subterranea*). It has many resemblances to the true groundnut (*see* p. 23) but differs from it as a food plant in that its seeds, with a low oil content, are not valued as oilseeds but only for their rich starch and protein content. It takes its name from the district of Bambarra in Mali, and probably originated in West Africa where it is occasionally found wild. The stems are very short and prostrate, and the leaves with their long stalks arise thickly from them, so that the general appearance of the plant is that of a close bunch of leaves rising from almost one point on the ground. The flowers are a pale yellow, and the plant buries its seed-pods in the ground like the groundnut. The rows are often earthed-up to ensure better coverage of the pods, which have a wrinkled surface and contain usually one but occasionally two seeds (2A). The seeds may be red, white, black, or mottled but they always have a white hilum. They are hard and need soaking, splitting, or pounding before cooking. The plant is little known outside tropical Africa and Madagascar, in most areas of which it is only a minor crop whose use is confined to local diets. It is particularly popular in Zambia.

Other Exotic Legumes. There are a number of other tropical legumes which are only of very minor or local importance as sources of food. One such group comprises certain species of *Phaseolus*, additional to those figured on pages 37 and 39, but which are also pulses with similar characteristics. *P. aconitifolius* is the moth bean of India; *P. trilobus*, known as pille-pesara, is a perennial also from India, and has been introduced into the Sudan; *P. angularis*, the adzuki bean, is mostly grown in Japan and China; *P. acutifolius*, the tepary bean, is cultivated in Mexico and Arizona and is specially useful in giving a quick though small yield on very little rainfall.

The cluster bean, *Cyamopsis psoraloides*, is a not unimportant crop, often interplanted with others, in northern India and West Pakistan. *Psophocarpus tetragonolobus*, sometimes known as Goa bean, has winged four-angled pods which are cooked whole in the same way as those of the asparagus pea (p. 43), and also has edible roots; it is grown in tropical Asia and occasionally in West Africa. *Lathyrus sativus*, the grass pea, is grown in India and the Middle East largely as a fodder crop, but the seeds are eaten by people especially in time of famine; excessive consumption of them can cause a paralysing disease known as 'lathyrism'. *Kerstingiella geocarpa*, sometimes called the Hausa groundnut, is rather similar to the Bambarra groundnut and is grown in some of the drier parts of West Africa. One other legume worth mentioning is the tamarind tree, *Tamarindus indica*, whose pods contain a brown pulp round the seeds which is eaten fresh and used in seasonings, curries, and drinks in India and Ceylon.

TWO-THIRDS LIFE SIZE   *SEEDS LIFE SIZE*

1 PIGEON PEA flowers   1A Seed pods   1B Seeds
2 BAMBARRA GROUNDNUT young plant   2A Pods and seeds

RUNNER BEANS AND FRENCH BEANS

1   Scarlet Runner (*Phaseolus coccineus*). This native of South America is a perennial, but is usually grown as an annual in the British Isles. If left in the ground, the rootstock may survive a mild winter in warm localities, but the plants are unlikely to produce a satisfactory crop. The Scarlet Runner is by far the most popular green bean in Britain, but on the Continent various kinds of French beans are generally preferred. The crop requires a moderately rich soil; the use of manure or compost is usually advocated, but excess of nitrogen should be avoided and if necessary balanced by the addition of appropriate quantities of phosphates and potash. According to the mildness of the district, seed may be sown from the middle of April to early June to produce successive crops from July to early October. Most runner beans are vigorous climbers, growing up to 10 feet high if unrestricted. It is usual to grow them on canes or strings 6 to 8 feet high, pinching out the tips of the leading shoots when they reach the top of the support.

The structure of the flower is typical of the Pea family (Papilionaceae). There are 4 free petals — a large upper standard, 2 lateral wings, and a boat-shaped keel. The petals are usually red, hence the name Scarlet Runner, but some cultivars have white, or red and white flowers. The bean pods vary from 8 to 24 inches in length, depending on varietal differences in colour — various shades of pink, spotted, or blotched with black, or sometimes almost entirely black. White-flowered varieties produce white seeds, e.g., 'Czar', which is one of the varieties favoured for quick freezing. 'Hammond's Dwarf Scarlet' is a true dwarf, non-climbing variety, introduced in 1961. 'Princeps' is of medium height, but early and heavy cropping. Other named varieties include 'Goliath', 'Kelvedon Marvel', 'Prizewinner', 'Streamline' and the red and white flowered 'Painted Lady'.

2-8   French or **Kidney Beans** (*Phaseolus vulgaris*). Probably of South American origin, this very variable tender annual has been long cultivated in many parts of the temperate, sub-tropical, and tropical zones. The numerous varieties are very diverse in the colour of their flowers and in the colour, shape, and size of their pods and seeds. There is considerable confusion in their nomenclature.

'Climbing Purple-podded Kidney Beans' (2). Amongst the most striking varieties are those climbing beans with purple flowers and purple pods and also usually purple stems. The beans are stringless and remain tender for longer than the string beans which are more usually grown. They lose their purple colour when cooked, turning green like most beans, and comparing very favourably in flavour and tenderness. Two examples are 'Blue Coco', which has a rather short, broad pod, and the 'Climbing Purple-podded Kidney Bean' ('Haricot à Cosse Violette') (2), described in 'The Vegetable Garden' by MM. Vilmorin-Andrieux, nearly a century ago.

'Pea Bean' (3). The short, broad pods and roundish seeds of this curious variety have given rise to the erroneous belief that it is a hybrid between a pea and a bean, but it is undoubtedly a true bean. It is a climber of moderate vigour.

'Canadian Wonder' (4) is an old-established example of what might generally be regarded as a typical French or Kidney bean, of dwarf or bush type of habit. It is a standard, mid-season commercial variety, included in the list of varieties given in the Ministry of Agriculture, Fisheries and Food Bulletin on 'Beans'. Other well-known, bush-type, green string beans include 'Masterpiece' and 'The Prince', both earlier than 'Canadian Wonder' and often used for forcing under glass.

'Deuil Fin Précoce' (5) is a dwarf French Bean, recently introduced, reputed to be early and very prolific. It has long, slender pods, green mottled with violet.

6-7 'Haricot Beans' are varieties of *Phaseolus vulgaris* which are grown in countries with warm climates primarily for their ripe seeds which, when dried, can be stored for long periods. These dried haricot beans are soaked in water over-night before being cooked. They are used in various dishes, two of the best known being Boston baked beans (U.S.A.) and *cassoulet* (France). 'Brown Dutch' and the white-seeded 'Comtesse de Chambord' are two well-known varieties which can be grown quite successfully in British gardens, although not considered worthwhile on a commercial scale.

8   'Mexican Black' is another unusual bean with yellow pods and black seeds which in flavour have been described as mushroom-like.

The 'wax-pod' group includes many yellow-podded stringless beans which are rarely grown commercially in Britain, although seeds are readily available. The stringless beans, both wax-pod and green, deserve to be better known, when both the runner and French beans commonly sold in our markets so often live up to their name 'string beans'.

PLANTS × ⅛   FLOWERS AND PODS × ⅔   SEEDS × 1   FLOWER SECTION × 2

1 SCARLET RUNNER   1A Flowers and pod   1B Seed
2 'CLIMBING PURPLE-PODDED KIDNEY BEAN'   2A Seeds
3 'PEA BEAN' pod   3A Seeds   4 'CANADIAN WONDER'   4A Flower section   4B Pod   4C Seed
5 'DEUIL FIN PRÉCOCE' flowers and pod   5A Seeds   6 BROWN HARICOT seeds
7 WHITE HARICOT seeds   8 'MEXICAN BLACK' flowers and pod   8A Seeds

37

TROPICAL PULSES

As the word 'pulse' implies, these are all leguminous plants which are grown to yield mature seeds for food, often used after drying and storage. The Indian word 'gram', meaning a pulse, is often used to designate some of these and other tropical legumes. All these crops are propagated by seed. Their special nutritive value is in the high protein content of the seeds.

1 **Butter Bean** (*Phaseolus lunatus*). Also known as the Lima or Madagascar bean, this plant has the largest seeds of this group. Originating in South America, it is now widely grown on a small scale in tropical and subtropical countries. It may behave in different climates as an annual or biennial. There are short erect varieties which can be grown without supports, and climbing ones which need poles to support them. Harvesting begins about 100 days from sowing, and may continue for some months, this extended period giving an unusually heavy yield for a pulse crop. White-seeded forms are usually grown, as those with coloured seeds have a rather high content of a poisonous substance which can be dangerous. A number of improved varieties have been developed in the United States. There is a small import trade in these beans to cooler countries where they cannot be grown.

2 **Chick Pea** (*Cicer arietinum*). It is important as the chief pulse crop of India, where it is known as Bengal gram and makes a large contribution of protein to the diet. It probably originated in western Asia. The leaves are divided into very small leaflets, making a light feathery foliage; the plant has a stout tap-root. The crop is ready for harvest 4 – 6 months after sowing. The small round pods contain one or two seeds each (2c). A variety called Kabuli gram, sometimes classed as a separate species, has larger seeds than the usual type. In India chick peas are usually grown as a cold weather crop. Before the seeds are cooked in the typical Indian pulse dish of 'dhal', they are split in a mill and separated from the husks. The crop is widely grown from northern India westwards all through the middle eastern countries; although of ancient cultivation in southern Europe and introduced into many

other parts of the world, its importance outside Asia is slight.

3 **Black Gram** (*Phaseolus mungo*) is often known by its Indian name of 'urd' and is called 'woolly pyrol' in the West Indies. It probably originated in India and is chiefly grown there, often in mixed cultivation with other crops; in the West Indies it is sometimes planted as a green manure. It is mainly useful for the dry seeds which are eaten as pulse, but young pods are also sometimes boiled and eaten. The whole plant is hairy and has an erect or somewhat trailing habit similar to the dwarf French bean. Each pod contains up to 10 black seeds which have a conspicuous white hilum (3B). The crop is ready to harvest for mature seed about 4 months after sowing.

4 **Green Gram** (*Phaseolus aureus*) is often known by the Indian name of 'mung' and is probably native to India. The plant is rather similar to black gram but it is somewhat less hairy. The flowers are purplish-yellow, and a pod may contain up to 15 seeds. The seeds are green, brown, or mottled. This crop is again most important in India, but is grown on a small scale in many parts of tropical and subtropical Asia, Africa, and America. Its chief use is as a pulse, but in China and the United States it is also used to produce bean sprouts (see text figure) which are popular in certain dishes. These are obtained by germinating the seeds in the dark until the sprouts reach the desired length. Like other sprouted cereals and legumes, these sprouts have a useful vitamin content, and in time of famine can be used as a source of vitamin C for the avoidance of scurvy.

PLANTS × ⅛   *FLOWERS AND FRUITS* × ⅔   *SEEDS* × 1

1 BUTTER BEAN (LIMA BEAN) plant   1A Flowering stem   1B Ripe pod   1C Seeds
2 CHICK PEA plant   2A Flowering stem   2B Fruiting stem   2C Seeds
3 BLACK GRAM flowers   3A Pods   3B Seeds   4 GREEN GRAM shoot   4A Pods   4B Seeds

39

1 **Broad Bean** (*Vicia faba*). One of the most ancient of all Old Word cultivated vegetables, broad beans have been found associated with Iron Age relics in various parts of Europe, including the British Isles. Prehistoric specimens are all small-seeded forms — even smaller than the 'Horse bean' or 'Tick bean' varieties grown as food for livestock in modern times. The bean was regarded as harmful by upper class Greeks and Romans, who believed that eating it would cloud their vision. In recent times, it has been shown that a haemolytic disorder called favism, common amongst Mediterranean peoples but rare elsewhere, may be caused by eating broad beans.

Broad beans for the table are usually harvested when almost fully grown and the freshly shelled beans are boiled, steamed, or casseroled. Large quantities are also grown for canning and quick-freezing. Young pods, picked when only 2 or 3 inches long, can be cooked whole or sliced, like French and runner beans. The Broad Bean belongs to the Pea family (Leguminosae). It is an erect, hardy annual, easily recognizable by its 4-ribbed stem, single or sparsely branched. The compound leaves are composed of a few large leaflets and bear large stipules at their base. The white, black-blotched flowers are borne in axillary clusters. There are two main groups of varieties: the 'Windsors', which have short pods containing about four, large, almost circular seeds; and the 'Longpods', which have about eight, more or less oblong, rather smaller seeds. This is an arbitrary classification as breeding has produced varieties which are intermediate in character.

The Broad Bean is very hardy. In many parts of the British Isles, it can be sown in autumn for picking in mid-June, and from January to April for a succession of July and August crops. Black fly can be a bad pest, especially on spring sown crops. It can be controlled by spraying or dusting with suitable insecticides, but there is a danger of useful pollinating insects being killed, which can result in low yields through inadequate cross-pollination.

2 **Jack Bean** (*Canavalia ensiformis*). This is a legume originating in the American tropics and now widely distributed in tropical countries as a minor crop. It is grown mainly as a green manure or fodder crop but is sometimes used as human food especially in times of scarcity. Propagated by seed, the plant is a robust, deep-rooted annual, producing masses of often trailing foliage. Both the whole young pods and the mature seeds (2B), which are white in colour and of the size of broad beans, can be cooked and eaten, but the seeds must be used with caution as they can be slightly poisonous. There are differing local varieties, some with an erect and some with a more trailing habit.

A related species, *Canavalia gladiata*, is often called the 'sword bean', though the names 'jack bean', 'sword bean' and 'horse bean' are used somewhat indiscriminately between the two species. This species has a perennial climbing habit and is easily distinguished from *C. ensiformis* by the more strongly curved pods and by the red, pink, or brown colour of the seeds, which can be used to provide food in the same way. It originated in the Old World and its cultivation is more or less limited to tropical Asia.

PLANTS × ⅛   *PODS, FLOWERS AND SEEDS* × ⅔   *SECTION* × 1

1 BROAD BEAN plant   1A Flowers   1B Flower section   1C Opened pod   1D Seed
2 JACK BEAN plant   2A Flower   2B Ripe pod   2C Seed

41

PEAS AND LENTILS

1 **Pea** (*Pisum sativum*). Probably a native of the Near East, the pea was long cultivated in that region and in southern Europe before it arrived in Britain, reputedly brought by the Romans. Field peas (var. *arvense*) are grown for feeding stock and for use as green manure as well as for their seeds which are eaten in the form of pea meal or as split peas. The hardy field peas differ from garden peas in having purple, or purple and lavender, flowers and grey or dun-coloured seeds, sometimes finely spotted. Garden peas (which are, of course, grown in fields as well as in gardens) have white flowers, and their ripe seeds are yellowish white or bluish green. In recent years, an increasing proportion of the crop has been grown for quick-freezing and canning, with a corresponding reduction in the acreage of peas picked green for the market, or allowed to ripen for use as dried peas.

Garden peas may be classified broadly into several groups. Marrowfats, with usually large, wrinkled seeds, form the largest group and the most important for quick-freezing and canning as well as for the fresh green pea market. The numerous cultivars, such as 'Onward', 'Senator', and 'Harrison's Glory', show considerable variation in height, pod-size, and other characteristics. Most marrowfat types are not very hardy, and are therefore not used for autumn or first-early sowing.

2- 'Sugar Peas', 'Mangetout' or 'Edible-podded Peas' have
2A tender pods which can be eaten whole when young, that is, when the pods are still flat. Various cultivars have white or coloured flowers.

3 'Petit Pois' are, as their name suggests, small-seeded forms of French provenance. They are considered to be the finest flavoured peas, if picked when young and eaten when fresh.

Other groups include Blue Round Peas, which are hardy cultivars, used for winter sowing and characterized by having rather small, round seeds; examples are 'Meteor', 'Foremost', and 'Feltham First'; and White Peas, which are used mainly as dried split peas and are not much grown in Britain.

Pisum sativum, which comprises all the varieties described above, is a glaucous green, climbing, annual with large, leaf-like stipules. Its leaves consist of 1 to 3 pairs of leaflets, often bearing whitish markings. The tendrils, by which the plant climbs, are thread-like modified leaflets towards the tip of the leaf. The flower, white or coloured, is typical of the pea-family, the perianth consisting of standard, wings, and keel. The pod, if allowed to mature, becomes dry and disperses the seeds by developing a spring-like tension, which causes its valves to twist apart suddenly, scattering the seeds.

4 **Asparagus Pea** (*Lotus tetragonolobus*). Another native of southern Europe, this plant is grown for its edible pods, which should be picked when they are about 1 inch long, as they later become stringy. The common name is rather fanciful, as they bear little resemblance to asparagus in either appearance or flavour, but they are quite palatable if steamed and dressed with butter. The plant is grown more as a curiosity than as an economic vegetable crop.

The Asparagus Pea is an annual with hairy, prostrate stems, 6 to 16 inches long. The greyish-green leaves have 3 broadly ovate leaflets. The flowers are papilionaceous, about $\frac{3}{4}$ inch long, of a rather unusual and attractive brownish-red colour (4A). The mature pods are 2 to 3 inches long, with 4 prominent, longitudinal ribs (*tetragonolobus* means 4-angled pod). The smooth, brown seeds are about $\frac{1}{8}$ inch across.

5 **Lentil** (*Lens culinaris*). This is one of the oldest leguminous crops, believed to have originated in the Near East or the Mediterranean region, and known to the ancient Egyptians and the Greeks. It is still grown commercially mainly in the regions where it was first cultivated, although it can be grown as far north as the British Isles, given a warm, light soil. The seeds are marketed when dried and are widely used in soups and stews. They have a high protein content and are more easily digestible than animal proteins. In Catholic countries they have long been appreciated as a food for use during Lent.

The Lentil is a much-branched annual, up to 18 inches high, with slender, angular stems. Its leaves are pinnate, with 4 to 7 pairs of more or less oval leaflets, about $\frac{1}{2}$ inch long. The small white flowers are papilionaceous with a large upper petal, two lateral petals, and two narrow petals between these. The flowers are succeeded by short, flattened pods, each with 1 or 2 seeds. The seeds are green, greenish-brown, or reddish, sometimes mottled. Their shape is indicated by the classical name of the genus, *Lens*, from which the English word 'lens' is taken. The split and de-husked lentils of commerce are usually orange or reddish.

TWO-THIRDS LIFE SIZE   *PLANTS* $\times \frac{1}{8}$

1 GARDEN PEA   1A Flower   1B Plant
2 'DWARF SUGAR PEA'   2A 'MANGETOUT' flower   3 'PETIT POIS'   3A Flower
4 ASPARAGUS PEA   4A Flower   5 LENTIL   5A Flowers   5B Plant

43

SOME OTHER PEA-LIKE PLANTS

1 **Cowpea** (*Vigna unguiculata*). It is an annual legume, originating in Africa where it may still be found wild, but now cultivated in many tropical areas and in the southern United States. It is grown in two distinct forms and for two different purposes. In Africa and America the form commonly grown is a short erect or weakly trailing plant whose seeds are allowed to mature and are then used in the dried condition. In south-eastern Asia, on the other hand, the forms grown by the Chinese farmers who mostly cultivate the crop are tall climbing varieties needing sticks to support them, and it is the tender young pods which are cooked and eaten. These climbing varieties have longer pods than the erect kind, up to 16 inches long, and it is one of these types which is shown in the illustration (1B). There are also forms with pods up to 3 feet in length, known as 'yard-long beans', which are sometimes classified as a distinct species *Vigna sesquipedalis* and are chiefly grown in the Far East.

In the erect varieties, which have the reputation of being slightly more drought-resistant than most legumes, the pods can be harvested at about three months from sowing, when the mature seeds can be shelled out. The climbing varieties can begin to be picked for young pods at about 6 – 8 weeks from sowing, and picking may continue for two months.

The nutritional value of cowpeas lies in the high protein content which they share with other legumes, whether it is the seeds or pods which are eaten. The erect forms are most important in Africa, and especially in West Africa where in the higher rainfall areas cowpeas are often the chief pulse crop. They are also an important crop in the West Indies; in the United States they are more grown as a forage and green manure crop than for human food. In Africa the leaves of the crop are sometimes cooked and eaten. The tender green pods of the climbing forms are appreciated as a vegetable by the inhabitants of all races in Malaysia and neighbouring countries.

A related species, *Vigna vexillata*, which grows wild and is occasionally cultivated in Ethiopia and the Sudan, has a swollen starchy root which is used for food.

2 **Lablab** (*Dolichos lablab*) is also sometimes called 'bonavist bean' and 'hyacinth bean', or known by its Arabic name of '*lubia*'. It is a legume of Asian origin, cultivated in India since very early times, and elsewhere now chiefly grown in Egypt, the Sudan, and south-east Asia. Although a perennial, it is usually treated as an annual in cultivation, and will mature its seeds in 3 – 4 months from sowing. As a field crop, the plants stand erect or trail over each other, but as a garden vegetable in India and Malaysia they are provided with sticks up which they climb. The flowers are white or purple, the pods can be of various colours including pale yellow, and the seeds may be white, reddish, black, or mottled. Whole young pods are boiled and eaten in India and Malaysia, especially in curry. For this purpose picking can begin three months from sowing and may in a suitable climate continue for a year or more. The ripe seeds are also cooked as food, in India as a split pulse; but they should not be eaten raw as this can cause poisoning. The foliage of the crop is a most valuable animal fodder which can also be made into hay; in the Sudan the crop, which is grown on a large scale in rotation with cotton, is mainly used for this purpose. The plant is very sensitive to day-length; there are varieties which will only flower when the days are at their shortest, and others which will do so in longer days, a factor which has to be taken into account when introducing seed into new areas.

A related species, *Dolichos biflorus*, is known as 'Madras gram' or 'horse gram'. It is grown in India and Ceylon mainly as an animal fodder or green manure, but is also eaten by people as a pulse.

APPLES (1): CRAB APPLES

The chief importance of the wild crab apple, *Malus pumila*, is that it is the parent species of our cultivated apples. The progeny of *M. pumila* is most variable and this variability has made improvement possible by selection from countless seedlings, most of which will themselves have been the product of one or more chosen parent.

The fruit peculiar to the apple tree and the pear tree is called a 'pome'. The receptacle, which surrounds the ovaries in the flower and which is surmounted by the calyx and stamens, enlarges during the summer to become edible and juicy and enclose the cells which contain the seeds. In the ripe fruit the remains of the calyx and stamens are known as the eye.

Crab apples grow wild in many regions of Europe, Asia and America. Most of the hedgerow crabs are, in fact, seedlings of cultivated sorts which have reverted to the wild apple, for it is only rarely that self-sown seedlings prove to be large and valuable fruits by a process of natural selection in competitive surroundings.

Crab apples are sometimes planted in gardens for their decorative quality and for their fruit which makes excellent jelly but is otherwise scarcely edible.

1   **Malus pumila.** This is the common form of the wild crab. It has at least two apparently distinct varieties which interbreed, *M. sylvestris* — the form most often seen wild in Northern Europe, which is thorny at times with sour shiny green fruit and smooth leaves — and *M. mitis*, only native in Southern and Eastern Europe, which has no thorns, leaves downy, at least on the underside, and sweeter, coloured fruit.

2   **'Transcendent'.** This variety of *M. pumila* has a shiny, pale yellow fruit with a pink flush.

3   **'John Downie',** another variety of *M. pumila*, is characterized by conical fruit coloured golden and orange.

4   **Malus baccata** is a species, found through East Asia to North China, which is very resistant to apple diseases, and which has interbred freely with other crab apples. The brilliant red cherry-like fruit, with no trace of a calyx when ripe, makes excellent jelly. The round-headed tree is usually 40 to 50 feet high. *M. manchurica* is a related form with larger fruit, and is the one most cultivated. Several other Eastern Asiatic species, including *M. hupehensis*, are also popular in gardens for their beautiful flowers.

5   **'Golden Hornet'** is easily recognised by its small, bright golden fruit.

Classification of Apples. Centuries of cultivation and selection have resulted in more than 3,000 named sorts of apples. They are recognizable by a combination of many characteristics: the height and habit of the tree (spreading or upright branches, etc); the shape, character, and colour of the leaf; the fertility of the tree — some are self-fertile, while others need pollinating, and some flower freely, others sparsely; the time of flowering; the quality, texture, size, shape, colour, flavour, and scent of the fruit, as well as its season, and its keeping qualities; and the tree's hardiness and disease resistance in different climates.

Many systems of classification have been tried but most of them are too artificial to be much help. The National Fruit Trials at Brogdale are based on a simple scheme originally proposed by Bunyard. His sections are:

i. The Lord Derby group. The smooth sour green apples, a few of which turn white or yellow when ripe. None of them are sweet and none show any striping, though some may have a brown flush in the sun.

ii. The Granny Smith group. The green sweetly flavoured apples, some of which have brown or orange flush.

iii. The Lane's Prince Albert group. The striped sour green apples — like the last group but striped red and flushed, and sometimes crimson all over like Crimson Bramley.

iv. The Peasgood group. Smooth, striped and sweet apples.

v. The Golden Noble group. Smooth golden apples, without stripes, sometimes brown or pink flushed.

vi. Worcester Pearmain group. A group of red-skinned fruit, some shiny and some rough.

vii. The Blenheim and Cox's or Reinette group. Apples with a dry or rough skin with some russet, with orange red or crimson flush or stripes.

viii. The Russet group. More or less russetted on a golden or green ground, without red.

TWO-THIRDS LIFE SIZE   SECTIONS × 1

WILD CRAB APPLES

1 MALUS PUMILA   1A Blossom   1B Sections of flower and immature fruit
2 'TRANSCENDENT'   3 'JOHN DOWNIE'
4 MALUS BACCATA   5 'GOLDEN HORNET'

Apples are the most important and most widely cultivated fruits of the temperate regions, successful in a wide range of climates and soils. They have been grown for at least 3,000 years, and are valued especially for their good keeping qualities; until the 20th century, hardly any other fresh fruit was available in winter. Names became firmly attached to certain fruits only in the 17th century. The earliest English apples were probably introduced from the mainland of Europe. From these many seedlings were raised. This process was repeated as America and Australia were settled, and newly raised apples later sent back to England.

2 *16th Century.* **'Court Pendu Plat'**, known for 400 years and possibly since Roman times, is a richly flavoured little dessert apple, unusual in its very late flowering season. (*December – April*)

'London Pippin' or **'Five Crown'**, possibly a continental sort, has been known in Somerset since 1580. It is a very late green dessert or cooking apple, now widely grown in Australia. (*February – March*)

'Royal Russet' was known before 1597. This large, late, cooking russet keeps well and has a good flavour. It is widely known on the continent. (*till March*)

'Golden Pippin' was the most highly prized dessert apple in the 16th and 17th centuries. It has a golden yellow skin with an orange flush and a remarkably crisp and juicy flesh. There are some eighteen different Golden Pippins. (*November – March*)

'Nonpareil', known before 1600, is a small conical apple with greenish rather softt bu crisp flesh. It is mentioned in 1870 as "so well known as to need no description". (*till March*)

'White Joaneting'. This very old sort, known before 1600, is still the earliest of all apples to ripen. It is a greasy, yellow-skinned fruit sometimes red-flushed, of pleasant flavour. (*July*)

17th Century. **'Autumn Pearmain'**, mentioned in 1629, is a typical example of the hardy, disease-resistant apples of the time. It has a golden yellow fruit with some russet. (*September – October*)

'Golden Reinette'. Known before 1650, these continental dessert apples of good flavour, are usually russetted and red-yellow. They are widely grown on the Continent. (*November – March*)

'Boston Russet'. Raised before 1650 in the U.S.A., this delicious apple probably comes from pips taken to America by emigrants. (*January – March*)

'Devonshire Quarrenden'. Known before 1650, it was possibly originally French. It has a deep crimson fruit with white juicy flesh. (*August – September*)

'Flower of Kent'. Popular for several hundred years, like many others now almost forgotten, this large green fruit is said to be the apple that Sir Isaac Newton observed falling. (*November – December*)

18th Century. **'Ashmead's Kernel'**. This apple, raised in 1720 by a Dr. Ashmead in Gloucester, was voted the best flavoured apple in a recent tasting conference. It is a poor cropper. (*December – March*)

'Newtown Pippin'. Known before 1760, this crisp juicy dessert apple is yellow or greenish-yellow. It is widely grown in the U.S.A. and Canada, but is not always successful in England. (*till March*)

'Hawthornden'. Raised before 1790 at Hawthornden in Scotland, this is an excellent cooking apple, with an almost white skin. (*October – December*)

'Wagener'. Raised in New York State before 1800, this is an example of many good apples raised in the U.S.A. from European stock. The rather hard fruit is golden with a carmine flesh. It is remarkably prolific and disease-free. (*November – February*)

'Cornish Gilliflower'. Known before 1800, this delicious late apple needs a warm climate. (*December – May*)

19th Century. **'Blenheim'**. Raised at Woodstock before 1818, this was for a hundred years the most prized winter apple. It is large, yellow-skinned with a dull red flush, and rather acid. The large, long-lived tree breeds fairly true from seed. (*November – January*)

3 **'Pitmaston Pine Apple'**. Raised at Pitmaston before 1820, this remarkable little fruit is crisp and juicy, but more aromatic than its supposed parent Golden Pippin (*December – January*)

4 **'Coe's Golden Drop'**. Known before 1820, long lost, but discovered in Essex by Gerald Finzi, this apple is greenish-yellow, with a very long stalk. (*November – March*)

'Rosemary Russet'. Recorded by 1830, this perfect russet probably existed long before. (*till February*)

'Tom Putt'. Known before 1840, this large crisp acid fruit with a vivid red-striped skin is disease free. (*till November*)

'White Transparent' Introduced before 1850, this apple is characteristic of Russian and Scandinavian types. The smooth, shiny, white-yellow fruit is digestible, crisp, and juicy. (*August*)

'American Mother'. Sent before 1850 to England from America, this delicious apple is striped red all over, conical, with a richly flavoured, juicy flesh. (*October – November*)

5 **'May Queen'**. Known before 1890, the tree is unlike any other apple in that it bears heavy crops of full-size fruit when under 18 inches high. (*till June*)

'Bismarck'. About 1890, this vivid crimson cooking apple was first sent to England from Tasmania. (*November – February*)

6 **'Christmas Pearmain'**. Introduced about 1895, this characteristic Pearmain is rather dry and hard, but keeps well. (*November – January*)

TWO-THIRDS LIFE SIZE
1 APPLE BLOSSOM ('MAY QUEEN')
APPLE VARIETIES
2 'COURT PENDU PLAT'   3 'PITMASTON PINE APPLE'
4 'COE'S GOLDEN DROP'   5 'MAY QUEEN'   6 'CHRISTMAS PEARMAIN'

49

APPLES (3): VARIETIES THROUGH THE SEASON

In their wild state apples ripen in late autumn before the onset of winter. Varieties originating in Northern countries where summer is short must ripen within ten weeks of flowering, but those in Mediterranean countries have six months between blossom time in March and leaf fall in November. By selection among native and foreign kinds we now have apples which ripen from July through till April and May the following year.

July – August. The first are often those of Russian or Scandinavian origin and probably all descend from a northern variety. They are shiny-skinned fruits with soft, juicy flesh, acid but often delicious in their season, as much used for dessert as for cooking. They are usually white or palest yellow with a brilliant crimson or scarlet flush or few stripes:

'Akero'
'Biela Borodowka'
'Duchess of Oldenburg'
'Red Astrachan'
'Scarlet Pimpernel'
'White Transparent'

Later in August and probably derived from sorts named in the previous group, are other dessert varieties. These are mostly brightly coloured, shiny- or greasy-skinned apples which do not keep for long:

'Beauty of Bath'
'Feltham'
'George Cave'
'Irish Peach'
'Lady Sudeley'
'Laxton's Advance'
'Miller's Seedling'
'Red Melba'

August – October. The Codlins ripen in August and September. They are soft, acid, white-fleshed cooking apples with pale whitish or yellow greasy skins, usually oblong or oval in shape with a distinct nose:

'Emneth Early' (1)
'Keswick Codlin'
'Lord Grosvenor'
'Lord Suffield'

By September and early October there is a wider range of dessert apples. Among these there are good, sharp-flavoured striped apples:

'James Grieve'
'Michaelmas Red'
'Tydeman's Early Worcester'
'Worcester Pearmain'

Ready now, too, are many derivatives of Cox's Orange:

'Ellison's Orange' (2)
'Laxton's Fortune'
'Merton Worcester'
'Sunset'

In September and October the cooking apples ripen that derive from the Codlins and mainly possess their pale green or golden greasy skin, with a flush:

'Golden Spire' (3)
'Arthur Turner'
'Charles Eyre'
'Grenadier'
'Rev. W. Wilks'
'Stirling Castle'

Late October – November. Now many of the Reinettes, Pearmains, and Pippins ripen — mostly smaller, drier more richly flavoured dessert apples — and some Russets, all harder, less perishable fruits than the early sorts:

'Autumn Pearmain'
'Egremont Russet'
'Golden Pippin'
'Herring's Pippin'
'King of the Pippins'
'Laxton's Reward'
'Mother'
'St. Edmund's Pippin'

December – January. There are innumerable sorts of the type just described, dry-skinned, russetted apples with hard, juicy, aromatic flesh:

'Braddick's Nonpariel' (4)
'Adams' Pearmain'
'Blenheim Orange'
'Claygate Pearmain'
'Cox's Orange'
'Ribston Pippin'
'Tydeman's Late Orange'
'Winston'

Less usual at this season are the shiny smooth-skinned fruits:

'Coe's Golden Drop'
'Golden Noble'
'Lombart's Calville'
'Spartan'

The late cooking apples are mainly shiny, green apples with striped, hard, acid fruits cooking to a froth, keeping until the spring if in good condition and by then quite eatable raw:

'Annie Elizabeth'
'Bramley'
'Crawley Beauty'
'Howgate Wonder'
'Lane's Prince Albert'
'Mère de Ménage'
'Newton Wonder'
'Tom Putt'

January – April. There are a few later varieties which stay hard and green until March and April. These are all green or flushed crimson, shiny when ripe:

'Edward VII' (5)
'Monarch' (6)
'Forge'
'French Crab'
'Gooseberry'
'Hormead Pearmain'
'London Pippin'
'Lemon Pippin'

A few late cooking apples are russetted:

'Diamond Jubilee'
'Reinette de Canada'
'Royal Russet'
'Woolbrook Russet'

The very late ripening dessert apples belong mainly to the Reinettes group, small, green or green yellow apples with hard, dry, but well flavoured flesh, sometimes flushed with crimson red or russetted:

'Allen's Everlasting'
'D'Arcy Spice'
'Heusgen's Golden Reinette'
'King's Acre Pippin'
'Laxton's Pearmain'
'Laxton's Royalty'
'Orleans Reinette'
'Rosemary Russet'
'Sturmer Pippin'
'Wagener'

A few are green with golden or orange tint:

'Easter Orange'
'Grange's Pearmain'
'Granny Smith'
'Ontario'
'Sanspareil'

'May Queen' is a vivid crimson.

These very late apples need a long ripening season and may in a cold year lack flavour and prove difficult to store. They are essentially fruits for a good soil and a warm climate. Sturmer Pippin, for instance, is much more successful in Australia than in England.

TWO-THIRDS LIFE SIZE

APPLE VARIETIES

1 'EMNETH EARLY'   2 'ELLISON'S ORANGE'

3 'GOLDEN SPIRE'   4 'BRADDICK'S NONPAREIL'

5 'EDWARD VII'   6 'MONARCH'

APPLES (4): VARIETIES OF FLAVOUR AND QUALITY

The complex of acidity, sweetness, and bitterness, together with the scent, makes up the flavour of the fruit. In quality the flesh may be soft and mealy, or crisp, or hard. The proportion of acidity to sweetness is a most noticeable feature. Cooking apples with a high acid content cook to a froth, and if sweetened have a good flavour. Many dessert apples have as much acid but also more sugar, and this balance makes for a richly flavoured fruit which also cooks well. Bitterness is chiefly prized in cider fruit but in a small degree it imparts a nut-like flavour to cooking and dessert apples.

1 **'Cox's Orange Pippin'** is perhaps the favourite English dessert apple, probably a seedling raised about 1850 from the still highly prized 'Ribston'. It ripens well in England and cooks well. Its acidity is nicely balanced by sweetness, the skin has a strong, delicious scent, and the flesh a crisp texture. Such perfection has led many raisers to use it as a parent for new and more vigorous sorts — for it is difficult to manage. Its notable descendants include 'Ellison's Orange' (p. 51).

2 **'Egremont Russet'** of unknown origin, is the best of the Russets, a group with a sweet strong taste, crisp and firm, with little juice though never tough or hard, less acid than 'Cox's Orange', and usually without much scent in the skin. It crops well as a small tree. 'St. Edmund's Pippin', 'Sam Young', 'Ross Nonpareil' and 'Boston Russet' all have the same pleasant qualities, and their roughened skin seems resistant to scab.

3 **'Golden Delicious'** is characteristic of an entirely different group, the shiny thin-skinned hot-climate apples which possess a refreshingly light flavour, a crisp flesh, and plentiful juice. Though not often good enough in Britain, it has become the leading apple in most warmer countries.

4 **'Orleans Reinette'.** Like 'Cox' and 'Ribston Pippin' this apple can be susceptible to disease in cold areas and on different soils. The best flavoured of the rough-skinned Reinette group, it is sweeter and less juicy than 'Cox's Orange'. It has been known since before 1776.

5 **'Calville Blanche d'Hiver'.** This very old French variety is often mentioned as being the most delicious of all apples. It can only be grown to perfection in warm soils or in a cool greenhouse. Its large golden fruit possesses a light sweet scent of surprising strength and a delicious sweet juicy flesh.

'Gravenstein' is often considered the best flavoured apple in Northern Germany and Denmark. It is a fairly large fruit, its strong sweetness balanced by a marked acidity and a penetrating scent. The large tree is hardy and disease free.

'Granges Pearmain' is representative of the green late apples of which 'Newtown Pippin' is the superb example. The latter apple was raised in the States some two hundred and fifty years ago and was being shipped to London as a delicacy as early as the mid-nineteenth century. These are crisp firm apples with a strong flavour somewhat reminiscent of the pineapple.

'American Mother' and a few other sorts have a peculiar and not at all unpleasing scent or flavour of anise.

'James Grieve' is one of the most acid dessert varieties but it is also accepted as one of the most delicious and refreshing. Its scent is almost as acid as a lemon. Others of this type, for those who like acid fruit, are 'Upton Pyne' and 'Herrings Pippin'.

Cultivation. In the garden, apples are most successful if grown as dwarf bushes, pyramids, or cordons. For this purpose they are budded on a dwarfing rootstock. The older types of rootstock have now been superseded by the new 'Malling 26' which produces compact fruitful trees cropping from their second season. Nurserymen supply trees on this stock. It is essential to plant the tree so that the union of stock and scion is well clear of the ground. If the union is set near the ground or buried, scion-rooting will take place and the effect of the dwarfing stock will be lost. Apples thrive in a wide range of soils but not in poor acid soils or heavy clay. It is difficult to grow the less hardy kinds in high rainfall areas or above the 800 feet contour in most counties. When conditions are difficult, trees grown on Crab stock, which is more vigorous and able to penetrate deeply, should be tried. These trees take longer to bear fruit than those on dwarfing stock but when they do bear they will crop heavily.

TWO-THIRDS LIFE SIZE

APPLE VARIETIES

1 'COX'S ORANGE PIPPIN'   2 'EGREMONT RUSSET'
3 'GOLDEN DELICIOUS'   4 'ORLEANS REINETTE'
5 'CALVILLE BLANCHE D'HIVER'

APPLES (5): MODERN VARIETIES

Much time and patience was devoted to the improvement of the apple during the nineteenth century by amateurs and professional gardeners who raised innumerable seedlings and tried out many found wild, the most famous of which was 'Cox's Orange Pippin' (p. 53). This was recognised as superior to any previously known sort, and was used as a parent for breeding from about 1880 on. In the twentieth century fruit research stations were set up in many countries. Some have, by now, a large breeding programme of several generations from which they produce new, superior varieties.

1 **'Laxton's Fortune'.** This cross between 'Cox's Orange' × 'Wealthy' was produced by Laxtons, the notable English nurserymen, in 1904. It is perhaps the best all-round September apple. Other early apples of similar parentage are 'Laxton's Advance', 'Laxton's Epicure', and 'Laxton's Exquisite'. Among later ripening apples, 'Laxton's Superb' is, after 'Cox's Orange', now the standard market sort. It is a heavy cropper and ripens over a long period. Other late varieties are 'Laxton's Rearguard', 'Laxton's Reward', and 'Laxton's Royalty', which have as their second parent, old, late-flowering, long-keeping sorts such as 'Court Pendu Plat' (*see* p. 49). Their late-flowering tendency is useful in places subject to late frosts.

2 **'Tydeman's Early Worcester'.** At East Malling research station crosses were made to improve the quality and lengthen the season of the two most important market sorts, 'Cox's Orange' and 'Worcester Pearmain'. This and the next variety are two of the resulting crosses. 'Tydeman's Early Worcester' is a cross between 'Worcester' and the Canadian variety 'McIntosh Red', improved in size and texture.

3 **'Tydeman's Late Orange'**, a cross like the preceding variety, is like a late 'Cox's Orange', and more disease resistant.

4 **'Merton Charm'.** One of a wide range of varieties from the John Innes Institute, this is an October apple, like a redder 'Cox's Orange'. Other notable John Innes apples include 'Merton Beauty', a large September – October apple, an improvement on 'Ellison's Orange'; 'Merton Joy' and 'Merton Prolific', both late keepers; 'Merton Russet', a late variety; 'Merton Worcester', similar to 'Worcester' in season, but flavoured like 'Cox's Orange'.

5 **'Spartan', 'Ida Red',** and **'Mutsu'** are examples of the many good fruits produced by overseas research stations. 'Spartan', raised in Canada, has soft, sweet, dark crimson fruit, and is very free-cropping and resistant to disease. It ripens in November. 'Ida Red' from the United States, is a seedling from the old variety 'Wagener', notable for its long-keeping qualities. 'Mutsu' from Japan, a large, green, smooth fruit, crops heavily in December – January.

'Exeter Cross'. At Long Ashton a series of early disease-resistant dessert apples mostly derived from 'Worcester' or 'Cox' have been raised. These range from 'Exeter Cross' and 'Cheddar Cross' which ripen in August, to 'Hereford Cross' and 'Taunton Cross' ripening in September, and 'Worcester' and 'Gloucester Cross' in October.

'Scarlet Pimpernel', 'George Cave', 'Discovery'. These are three new early apples, interesting in being produced by amateurs, who still provide many of our best sorts, either as crosses or as chance seedlings.

Modern Orchards. The regions where apples are main crops include Europe, Lebanon and other near-Eastern countries, United States, Canada, Chile, Argentine, Australia, and New Zealand. Standard trees have been supplanted by bush or other dwarf forms. The grass is kept short by mowing so that it does not deprive the shallow tree roots of moisture or nourishment.

The sort most generally grown in Britain is 'Cox's Orange'. For earlier picking 'Scarlet Pimpernel', 'George Cave', 'James Grieve', 'Worcester Pearmain' and its crosses, 'Laxton's Fortune', and 'Sunset' are all grown. For later fruit, 'Laxton's Superb' is most popular, but to extend the season with the aid of cold storage other sorts are being tried such as 'Duke of Devonshire', 'Tydeman's Late Orange', and 'Golden Delicious'.

In countries with a hotter summer, the typical shiny apples of warmer climates are now grown almost to the the exclusion of other sorts. 'Golden Delicious' is the leading variety (*see* p. 53).

Other well known sorts include 'McIntosh' and its improved forms such as 'Cortland'; 'Democrat' (a very dark crimson Australian apple); 'Melba', and 'Jonathan' (red Canadian sorts); 'Rome Beauty', and its several derivatives, all red; 'Newtown Pippin'; and 'Granny Smith'.

TWO-THIRDS LIFE SIZE

APPLE VARIETIES

1 'LAXTON'S FORTUNE'   2 'TYDEMAN'S EARLY WORCESTER'
3 'TYDEMAN'S LATE ORANGE'
4 'MERTON CHARM'   5 'SPARTAN'

PEARS (1)

Though much like the apple in many respects, the Pear usually has a more elongated fruit, gritty to the taste, and a thicker, fleshier stalk not set deep in the fruit. The Pear belongs to the genus Pyrus which is distributed across the warmer parts of Europe, Asia Minor, India and North-East Asia. Cultivated pears are all descended from the Common Pear, *P. communis*.

1 P. communis is a handsome, lofty tree which may be found growing apparently wild over much of Europe, but more in the South than the North. The wild pears have a tendency to be spiny when young and their fruit is usually hard and green. Pear timber, which is known as fruit wood, is much sought after for furniture making, having an even grain and a pinkish-brown colour.

There are several other European wild Pears: *P. cordata* is a small tree, native to France, Spain, and Portugal, with small brown fruit; *P. amygdaliformis* has narrower leaves and grows on the Northern Mediterranean shore; *P. nivalis* and *P. salicifolia* come from Eastern Europe and are often grown in gardens for their silvery foliage and white flowers.

The Chinese Sand Pear, *P. sinensis*, is a native of China with rather insipid fruit. American hybrids of *P. sinensis* and *P. communis*, resistant to disease, are known as Kieffer Pears.

2 'Hazel' or 'Hessle' is characteristic of many hardy old pears, of no value for market but very hardy and bearing immense crops of small juicy fruit.

3 'Fertility', typical of the older market pears grown near London, is a hardy tree which crops regularly, producing pears of good size and juicy, but now supplanted by better fruits.

4 'Conference'. The most widely grown pear in Britain, raised in Berkshire about 1770, it is resistant to scab and prolific. Though not self fertile, it will set a crop of fruit parthenocarpically without pollination.

5 'Williams' Bon Chrétien' or 'Bartlett' which ripens in late September, is perhaps the best known pear of all. This delicious fruit, very juicy with a rich muscatel flavour, is widely grown for canning in the U.S.A. where it is known as the 'Bartlett' pear, after the man who first took it there. It is unfortunately susceptible to scab.

6 'Glow Red Williams'. This is a mutant or bud sport with a strong red colouring in the skin of the fruit, the leaves, and the shoots. This red colouring in the skin makes it more resistant to weather and fungus damage. Three newer crosses with 'Williams' as a parent are 'Gorham', an American variety from 'Joséphine de Malines'; 'Merton Pride', a John Innes variety from 'Glou Morceau'; and 'Bristol Cross', from 'Conference'. These are all September ripening.

'Doyenné d'Eté, the first sort to ripen, is a delicious small fruit raised before 1700, which grows on an exceptionally weak, small tree.

'Jargonelle', another old sort (1600), follows soon afterwards in early August. This is a large spreading tree often seen in old gardens.

'Clapp's Favourite', 'Beurré d'Amanlis', and 'Dr. Jules Guyot' are three good prolific pears which ripen in early September.

'Marguerite Marillat' is remarkable for enormous crops of very large fruit borne on small upright branches.

Classification. Cultivated Pears vary a great deal in their habit of growth (erect or pendulous), in size, in flowering time, and in the amount of fruit they bear. All are more or less sterile and require pollination. They may live 200 or even 300 years. The fruits are usually classified into six shapes:

Round or flattened: These are mostly old varieties, with a rough skin and russetted.

Bergamot or top-shaped: Mostly rough-skinned varieties, russetted, or green when ripe.

Conical pears, tapering but not waisted: usually yellow, russetted, or red flushed.

Pyriform with a distinct waist: usually yellow when ripe, and they include many of our choicest pears.

Oval: a small group, mostly russetted without red colouring.

Calebasse (*long pears*). They are often brown or golden russet, but a few are smooth green, and these mostly cooking pears.

The fruit may be described as for dessert or stewing. Stewing pears, not acid but hard, lacking in flavour and juice, are mostly hardy, prolific, and long-keeping. They need slow cooking in a good syrup. They include: 'Bellissime d'Hiver', Beurré Clairgeau', 'Blackwater', 'Catillac', 'King Edward', 'Pitmaston Duchess', 'Uvedale St. Germain', 'Vicar of Winkfield'. The perfect dessert pear is juicy or buttery, acid yet sweet, with a strong delicious scent. Pears are more difficult than apples to ripen properly if the conditions of soil and weather are not favourable, and faulty ripening produces a gritty, tasteless, juiceless fruit which is hard before it is ripe, and mealy when ripe.

Pears vary in their season and remain perfect for a relatively short time. They should be picked just before they are ripe and stored in temperate conditions entirely apart from other fruit.

TWO-THIRDS LIFE SIZE · SECTIONS × 1

1 PEAR BLOSSOM   1A Sections of flower and immature fruit
PEAR VARIETIES
2 'HAZEL'   3 'FERTILITY'   4 'CONFERENCE'
5 'WILLIAMS' BON CHRÉTIEN'   6 'GLOW RED WILLIAMS'

PEARS (1)

In Europe, the centre of pear growing has always been in France, Belgium, and West Germany, a region with the required combination of sufficient moisture and summer warmth. In Belgium and France from 1775 onwards a few wealthy and painstaking amateurs produced the excellent varieties we grow today. It seems more difficult to raise good new pears than apples, possibly because there are fewer good parents to choose from.

1 **'Bergamotte d'Espéren'**, an old variety, is a delicious pear of excellent quality, sometimes ripe in December but keeping till March.

2 **'Packham's Triumph'** is one of the most recently introduced sorts, ripening in late October.

'Beurré Hardy'. This variety, like 'Louise Bonne of Jersey' ripens before the end of October. Both are large, vigorous trees. 'Beurré Superfin' produces choicer fruit but is more difficult to grow and to ripen properly.

'Doyenné du Comice' is the best November pear, and the one most widely grown in England. In favourable conditions it produces good crops of delicious fruit. It is pollinated by 'Marie Louise' — itself a good pear of this season — or by 'Glou Morceau', mentioned below.

3 **'Durondeau'** is characteristic of the russet-coated pears, a good quality fruit much grown at one time for market. 'Beurré Brown' also has a golden or reddish russet skin. This is one of the oldest sorts grown and the first sort to have its fruit described as 'beurré' in reference to the buttery texture of its flesh.

'Thompson's', though given an English name, is one of Van Mons' pears raised about 1810. It is one of the choicest garden pears.

'Joséphine de Malines' and **'Glou Morceau'**. These are the two latest-ripening pears which will grow well in Britain, ripening successively and bearing fruits of the highest quality.

4 **'Olivier de Serres'**. This variety, and the two succeeding, ripen very late, in March or April. Such late ripening pears succeed well in the long summers typical of southern France and Italy, but are difficult to grow in more northerly regions.

5 **'Passe Crasanne'**, ripening in March or April, is a russet fruit, much grown in Italy as an orchard tree but not ripening easily in England.

'Easter Beurré', is the last pear to ripen, a hardy fruit storing well with the same rich flavour as 'Williams' when ripe.

History. Some of the old sorts may still be seen in orchards. The early sorts 'Doyenné d'Été' (1700) and 'Jargonelle' (1600) have no equal in their season. 'Bellissime d'Hiver', 'Black Worcester', 'Catillac' and 'St. Germain' are old stewing pears. Characteristic of the hardy North country pears are 'Achan', 'Chalk', and 'Hessle'. Others sometimes encountered are 'Bergamotte d'Automne', which like 'Black Worcester' is possibly Roman, 'Beurré Brown', 'Bishop's Thumb', 'Forelle', 'Girogile', 'Lammas', 'Oignon', 'Swan's Egg', and the unusual 'Sanguinole' which has a red flesh as well as a red skin.

The fruit tree breeders of the 18th and 19th centuries in England produced only two outstandingly successful pears, namely 'Williams' Bon Chrétien' and 'Conference' (*see* p. 57).

From America came 'Clapp's Favourite', and the remarkable 'Seckle' discovered by a trapper about 1765, as a seedling, in woodland which he subsequently purchased. The Victorian growers raised many pears, among which the following survive: 'Pitmaston Duchess', raised by Williams; Thomas Andrew Knight's 'Althorp Crasanne', 'Broompark', 'Eyewood Bergamot', 'Monarch' and 'Rous Lench'; Thomas Rivers' 'Conference', 'Fertility', 'Magnate Princess', and 'St. Swithin'. The most widely grown of Laxton's varieties, 'Laxton's Superb', is now being dug up because it is susceptible to fire blight. Others of theirs include 'Early Market', 'Satisfaction', and 'Foremost'. Of the Victorian amateur raisers, the Rev. J. Huyshe of Clyst Hidon had some success with 'Prince Consort', 'Victoria' and 'Prince of Wales'.

In recent years from Long Ashton we have 'Cheltenham Cross' and Bristol Cross'; and from the John Innes Institute, 'Merton Pride'.

Cultivation. Until about 1870 most pears were budded on pear stock. This gives large vigorous trees which may take a long time to crop. Early cropping was often induced by periodic lifting and root cutting. Today most pears are grown on quince rootstock which requires a fairly moist, easily worked soil and will not succeed in the hard stiff soils that pear stock often thrives in.

Pears may be readily grown as bushes or pyramid trees in most gardens or as fan-trained or cordon trees on sunny walls in colder areas. As they flower several weeks earlier than the apple, they need warmer conditions earlier in the year. They also need a richer, moister soil than suits the apple, and in too alkaline soils they may show chlorosis. In England the best pear growing districts are in Kent where the climate is warm in summer and the best soils are a rich deep brick earth. Pears, like plums, often fail to crop because they are insufficiently pollinated — either the weather is unfavourable for the insects, or trees suitable for cross-pollination are not available. During late October and November the main crop of pears will ripen in England. This is perhaps a month later than they would ripen in France and Italy.

TWO-THIRDS LIFE SIZE

PEAR VARIETIES
1 'BERGAMOTTE D'ESPÉREN'   2 'PACKHAM'S TRIUMPH'
3 'DURONDEAU'   4 'OLIVIER DE SERRES'   5 'PASSE CRASANNE'   59

CIDER APPLES AND PERRY PEARS

The cultivated varieties of these fruits are descended from the same wild stock as the dessert and cooking varieties (*see* p. 46). They may be recognised by their bitterness or astringency, which comes from the tannins in their juice. The juice of apples and pears will ferment, by the action of yeast, to produce alcohol about twenty-four hours after it is expressed, unless fermentation is prevented, for example by pasteurisation. In America unfermented apple-juice is called 'soft cider', and fermented is 'hard cider'. The distilled liquor from cider—sometimes called 'apple brandy'—is known as 'calvados' in Normandy, and 'applejack' in America. Varieties with the requisite high tannin content have been grown for cider and perry for at least 2,500 years. In England cider apples were probably grown in Saxon times but perry pears may not have been introduced until the Norman Conquest, Normandy then as now being a great perry area.

1 **'Sweet Coppin' Apple.** This Devonshire apple with a low acid content, is classified as 'sweet', which is one of the three categories into which cider apples are divided, the other two being 'sharp' (with a high acid content); and 'bittersweet' or 'bittersharp' which are respectively sweet and sharp with a high tannin content. Usually in cider-making two or more varieties are blended together, since few varieties have all three characteristics of bitterness, sharpness, and sweetness necessary for making good cider.

2 **'Tremlitt's Bitter' Apple.** This apple, which is classified for blending purposes as 'bittersweet', is also a Devonshire fruit. Traditionally, the South West and the West Midlands are the areas of England most suited to cider orchards. Apples flourish in the Wye Valley and the Severn Vale, and the apple is the predominant orchard tree in East Cornwall, Devon, Somerset, Dorset, and Wiltshire.

3 **'Kingston Black' Apple.** This variety, classified as 'sharp', is considered the best of the cider apples, and is one of the few kinds which does not need to be blended with other varieties but makes an excellent single cider. It is, however, difficult to grow well. The English counties where it thrives include Somerset, Gloucester, and Hereford.

Cultivation. Most cider apples and perry pears have no recorded history and little can be learned of their origins except through their names which may be that of a farm, a village, or their raiser — though often they have simple names such as 'Golden Ball', 'Red Pear', or 'Brown Snout'. Farm orchards used to be planted in grass with seedlings raised from local trees; but the demand from scientifically controlled factories for fruit of uniformly good quality has in this century brought into being a new type of orchard with large numbers of trees, carefully arranged to cross-pollinate and to produce a succession of well balanced fruits. The Long Ashton Research Station helped in this orchard reform, and also made a survey of apple names (1939).

4 **'Yellow Huffcap' Pear.** This variety, one of the large family of 'Huffcaps', is an immense tree, even by comparison with other pears, which are often large.

5 **'Thorn' Pear.** This old variety, known in 1676, is unusually upright in growth.

6 **'Taynton Squash' Pear.** This old variety is recorded in Thomas Andrew Knight's *Pomona Herefordensis* (1811), where many varieties are identified which might otherwise have been forgotten.

7 **'Red Pear'** is another very old variety, known in Herefordshire and Worcestershire since Tudor times.

Cultivation. The pear juice for perry is generally not blended but taken from a single variety. The quality and composition of the juice from any one variety can vary a great deal in different soils and in different years, and experience is needed to produce good perry. The long-lived pear, with its powerful root system, seems to thrive in soils too hard and poor to support an apple, but it needs dry, sunny conditions. In orchards pears must be spaced widely so that the fruit may be more exposed to sunlight. The centres for pear culture in Britain are the Wye Valley and the Severn Vale, and in the last decade perry orchards have been planted in Somerset, with bush as well as standard trees. In 1963 a survey of pear names was completed by Long Ashton Research Station.

TWO-THIRDS LIFE SIZE
CIDER APPLES
1 'SWEET COPPIN'   2 'TREMLITT'S BITTER'   3 'KINGSTON BLACK'
PERRY PEARS
4 'YELLOW HUFFCAP'   5 'THORN'   6 'TAYNTON SQUASH'   7 'RED PEAR'

QUINCE, MEDLAR, ROSE, AZAROLE

These trees of temperate climates are all members of the Roseaceae family which bear fruits of some local interest but of no commercial importance.

1 Quince (*Cydonia vulgaris*). Probably a native of western Asia where it still grows wild, the quince has been cultivated since ancient times, and was valued by the Romans. The tree is small (15 – 20 feet), sometimes thickly branched and bent as if deformed by wind and the weight of its foliage, but often upright with strong branches. It has rounded oval leaves which are very woolly beneath, and solitary pink and white flowers (1A) at the ends of short young shoots in May. Its fruit is similar to that of the pear and the apple, but has many ovules in each carpel or section — up to twenty, instead of two. It is hard and acid but when cooked with sugar it turns a dull pink colour and makes a delicious jam or jelly. It is also valued as a flavouring to be added to cooked apples or pears. 'Contignac' is a candy made of quince purée and sugar. (The name is derived from the French name for this fruit, *coign*.) *Marmelo*, its Spanish name, is the origin of the term 'marmalade' used in English only of preserves made of citrus fruits. Besides being grown for its fruit, the quince is much grown as a rootstock for pears (*see* p. 59).

The usual quince varieties seen in England are: 'Bereczki', a Serbian variety lovely in flower; 'Champion', fertile, and with a comparatively mild taste; 'Ispahan', a very vigorous form collected in Persia; 'Maliformis', the apple-shaped variety; 'Meech's Prolific', noteworthy for its precocity in bearing; and 'Portugal', a choice but somewhat difficult kind.

Japonicas, now classed as Chaenomeles, are the familiar flowering shrubs closely related to the quince, but distinguished from the true Cydonia by their longer serrated leaves and united, not free, styles. They have rounded smooth fruits, yellow in *C. japonica*, white in *C. lagenaria*, and green in the uncommon *C. cathayensis* — the last often up to 6 inches long. These fruits are even more tart than the true quinces and when combined with apple or pear impart a strong, pleasant flavour to jam or jelly.

2 Medlar (*Mespilus germanica*). Hardier than the quince and able to naturalise in colder climates, it often appears wild in Britain and on the European continent. The medlar fruit is remarkable in that the five seed vessels are visible in the eye of the fruit, for the fruit is set in the receptacle as in a gaping cup, around the rim of which stand the five conspicuous calyx lobes. In countries like Italy where the fruit ripens it can be eaten off the tree. In colder countries the medlar does not become palatable until it is half rotten, or 'bletted', when it becomes soft, brown, and more palatable than might be expected. It also makes a pleasant jam if the seeds are removed. Of the sorts grown, 'Nottingham' is a small fruit, 'Royal' rather larger, and the 'Dutch' or 'Monstrous' the largest. The size of the tree is proportionate to that of the fruit.

The medlar is a spreading tree, apt to be deformed by the wind. The wild tree has thorns but the cultivated kinds are thornless. Its flowers are borne at the end of short young shoots in late May or June. The medlar may be propagated by grafting it on thorn, quince, or pear stock. Thorn trees on which branches of five very different trees have been grown are sometimes seen; these are the pear, the mountain ash or rowan, the whitebeam, the medlar, and the azarole.

3-4 Dog Rose (*Rosa canina*) and **Rosa rugosa.** The rose hip is an urnshaped receptacle, almost closed at the mouth, with numerous achenes enclosed within it. Roses are not cultivated for their fruit, but rose hips are now extensively collected from wild plants and used to make rose-hip syrup, rich in vitamins. Country people, especially in Northern countries, have always used rose hips for making jellies, preserves, and sauces.

5 Azarole (*Crataegus azarolus*). This Southern European member of the hawthorn genus is grown in France, Italy, Algeria, and Spain, where its fruits are used for making jam and as a flavouring in liqueurs. In England, where it has been known since the seventeenth century, it is often planted as an ornament. Its fruit, which has a distinct apple-like flavour, is like that of the hawthorn but larger, consisting of a few nut-like carpels embedded in a rather mealy flesh. The typical form has orange-yellow haws but there are also ivory white and darker red varieties.

Service Tree (*Sorbus domestica*). This relative of the rowan has clusters of hard acid fruit, rather like small brownish or green apples or pears, which are edible only when bletted. This is a native of Asia Minor which has been grown over most of Europe in the past.

Rowan Tree (*S. aucuparia*). The bright red clusters of fruit from the rowan, or mountain ash, are collected in Northern countries of Europe to make jelly and to flavour apple preserves.

TWO-THIRDS LIFE SIZE

1 QUINCE   1A Blossom   2 MEDLAR   2A Blossom

3 DOG ROSE   3A Section of flower   4 ROSA RUGOSA   5 AZAROLE   5A Flowers

CHERRIES

Cherries belong to the genus Prunus which also includes Plums (pp. 67 – 71) Peaches and Apricots (p. 73), and Bird cherries. Prunus is readily distinguished from related genera by its fruit which has thin-skinned juicy flesh enclosing a single stone and is described as a 'one-seeded drupe'. Prunus stones and sometimes the leaves too have a flavour of 'almond essence'.

The Cherries are distinguished by their shiny fruit borne in clusters on relatively long flower stalks or pedicels. Their white or pinkish flowers are produced in clusters.

1 **Sour Cherries** derive from *Prunus cerasus*. Morello (1), (sometimes known as *P. acida*) is the best of the sour cherries. It is obviously related to *P. avium* but is a much smaller tree with drooping branchlets, green young foliage not tinted brown, and shorter stalked fruit more acid than sour. The typical Morello has dark fruit. Wild or native Morello Cherries are used widely for making liqueurs.

The red sour cherries such as 'Flemish' and 'Kentish Red' are sometimes called red Morellos, and sometimes 'Amarelles' or 'Griottes'. They are similar to the typical Morellos in flower and leaf, but have light to medium red fruits, and are more tree-like with a good trunk. The Morello Cherry and its related varieties are relatively free from diseases and insect pests. The Morello itself grows into a pleasant small bushy tree or it may be grown on a North wall as a fan trained tree.

2-3 **Sweet Cherries.** The cultivated varieties arose from the Mazzard or Gean (*Prunus avium*), probably originally from an Eastern form of this tree from Asia Minor. The wild Mazzard is a tall and splendid tree up to 75 feet in height with white flowers, brown tinted young foliage, and a smooth silvery bark. It grows freely in well-drained woodlands and has a notably wider range than the orchard cherries, growing well in both acid and basic formations and in heavy rainfall areas. The sweet cherry trees closely resemble *P. avium*, but they vary greatly in stature and tree form, and will not thrive in wet areas. The wild Mazzard has small dark red fruits which are either sweet or bitter, but not acid. There are cultivated Mazzard cherries seen in Devon and other districts where the better sorts fail. They are all small, black, and richly flavoured.

Cultivated cherries are either dark red-black or pale yellow more or less covered with a red flush. Formerly they were classed as either 'Bigarreau' with firm crisp fruit, or 'Geans' or 'Guignes' with soft juicy fruit; but there are now so many intermediate forms that this division is hard to apply.

Many cherry varieties are of great antiquity. Some indeed seem unchanged since Roman times but raisers since the early nineteenth century have introduced many good new sorts, in Britain, on the Continent, and in America. Knight raised a large number of good sorts and in this century the John Innes Institute raised the Merton varieties. Some of the best varieties are:

Early white	'Frogmore Early'
Early black	'Early Rivers' (2)
	'Bigarreau Schrecken'
	'Merton Favourite'
	'Merton Heart'
Midseason white	'Kent Bigarreau'
	'Napoleon' (3)
Midseason black	'Roundel Heart'
	'Waterloo'
	'Merton Bigarreau'
	'Merton Bounty'
	'Merton Premier'
Late white	'Florence'
Late black	'Hedelfingen'

Sweet cherries are not only fastidious as to the situation and the climate, but the soil must also be satisfactory. In Britain growing is mainly limited to a very few favoured areas in Kent, Buckingham, Berkshire, Gloucester, and Hereford. In the open orchard they are plagued with more diseases than most fruit trees. The most disastrous and least possible to check is bacterial canker which will kill the tree. They are also subject to virus diseases, and virus-free stocks are essential.

As the Sweet Cherry grows into too large a tree for a small garden to accommodate, gardeners sometimes grow it in a pot — a method which checks the roots but often produces good crops.

Hybrids between *P. avium* and *P. cerasus* are a group valued for their hardiness and cooking qualities. These may also be either black fruited or red fruited and they are generally known as 'Dukes' or 'Royal Cherries'.

Bird Cherries (*P. padus*). These are hardy Northern trees with white flowers in racemes which appear on the young shoots in May. In the U.S.A. one species, *P. serotina*, at least provides fruit which is used for flavouring rum and brandy, for which purpose it is said to equal the morello cherry.

TWO-THIRDS LIFE SIZE   SECTIONS LIFE SIZE

1 MORELLO CHERRY   1A Blossom
SWEET CHERRY VARIETIES
2 'EARLY RIVERS'   2A Blossom   2B Sections of flower and immature fruit
3 'NAPOLEON'

PLUMS (1): SLOE, BULLACE, DAMSON, GAGE

There are numerous species of Plum which grow wild in the Northern hemisphere, from the Pacific Coast of America in the West to Japan in the East. They are slender deciduous trees of irregular growth with small white flowers in early spring and short-stalked fruit, borne usually singly or in twos or threes. Whilst the kinds grown in Britain are derived from the European *Prunus spinosa* and the Western Asian *P. cerasifera*, those grown in warmer countries are chiefly derived from the Japanese Plum *P. triflora*.

1 **Blackthorn** or **Sloe** (*P. spinosa*). This is the Wild Plum of Western Europe, a small tree, never much over 10 feet in height. Its small round blackish fruit have a blue waxy bloom and green, acid-flavoured flesh. It has thorny shoots which are dark purple and downy when young, and black bark. It has no value except for making sloe wine and sloe gin.

2 **Bullace** (*P. insititia*). This is a round, blue-black fruit much like the sloe in appearance but larger. Its flesh, though softer and more palatable than the sloe, is still so sharp that the fruit is usually left on the tree until November, long after the leaves have fallen, until frost has softened its acidity. The tree, less thorny and larger than the sloe, is seldom planted, but is plentiful in some places in a natural state. The best known form is the old 'Black Bullace', the fruits of which are more oval and purple-tinted than typical bullaces. The 'Shepherd's Bullace' (2) is another old kind with greenish-yellow fruit, but is rarely seen today.

3-4 **Damsons** (*P. damascena*). The Damson is a closely related fruit but generally it is more oval in shape, with less bloom, ripening at least six weeks before the bullaces and possessing an entirely different richer and sweeter flavour. It is cooked when ripe, and is also used for jam.

The 'Farleigh Damson' (3) is one of the best known varieties. In its native Kent it bears enormous crops and as it makes a sturdier tree than most, it is much planted as a windbreak.

The 'Prune' or 'Shropshire Damson' (4) is an old variety found in many West Midlands fields and orchards, growing half wild in hedges. It is never a heavy cropper but it is valued for its rich flavour.

The 'Merryweather Damson' is a larger fruit but of true damson flavour. These bullaces and damsons are easily grown in the garden, where they look best as small standard trees, but only flower and crop well in districts that suit them.

5-6 **Gages** (sometimes called *P. italica*). Of very different appearance from damsons, yet obviously closely related, are the gages. The typical greengage (5), which is found in an apparently wild state in Asia Minor, is a sturdy, rarely thorny bush or small tree with round green, deeply sutured, often red-spotted or russetted fruit with a slight bloom and firm green flesh deliciously scented and sweet flavoured. Though it may well have been cultivated in earlier times, it was re-introduced to England about 1725 from the Continent where it was known as 'Reine Claude', and took its name from its introducer, Sir Thomas Gage. The so-called 'Old Greengage' is probably this same fruit or a seedling from it — for it breeds fairly true from seed and may have been reproduced from seed regularly over a period of time without loss. 'Cambridge Gage' is one of these seedlings, now grown on a large scale for jam and canning, and said to be more reliable.

Some seedlings from the Greengage have larger fruit and though they may have a plum as one parent, they are still classed as Gages. These include 'Bryanston' and 'Reine Claude de Bavay'. 'Denniston's Superb' is an American variety, very near the greengage and almost equal in flavour, with the great advantage of being hardy, self-fertile, and a regular cropper. A new variety of Canadian origin is 'Ontario', the hardiest of the greengage type plums with a large fruit of good quality.

The original 'Transparent Gage' (6) is an ancient French variety, with golden almost translucent fruit with a light bloom and red mottling. It is perhaps the best flavoured of all plums, possessing a delicious honeyed sweetness. 'Early Transparent' and 'Late Transparent' were raised from stones of this sort by Rivers. The 'Mirabelle' is similar in appearance, but less hardy and therefore not often seen in England. The fruit has a rather sharper flavour and is famous for the apricot-like jam made from it. These golden gages mostly ripen towards the end of August. 'Denniston's Superb' and 'Early Transparent' ripen a fortnight earlier, while 'Late Transparent', 'Bryanston' and 'Reine Claude de Bavay' should be a fortnight later in mid-September.

Halfway between the plums and gages are several large round golden-fruited sorts, which are gage-like in flavour but plum-like in size and colouring. Of these the choicest is 'Golden Transparent', one of Rivers' raising, a large round yellow fruit ripening in October. It is self-fertile but not hardy except on a warm wall. 'Washington' is an old American variety, round and yellow with a pinkish red flush, ripening early September. Although difficult in cold areas, it is grown for market in others. To obtain a hardier type of gage, several breeders have raised crosses between gages and plums. Most successful have been Laxtons, with 'Laxton's Gage', and 'Laxton's Supreme'. Apart from the few kinds of known hardiness, gages in Britain are best grown on a sunny wall as fan-trained trees except in the area where they are known to thrive in the open — chiefly in the Eastern Counties, Kent, Essex, Suffolk, Norfolk and Cambridge.

TWO-THIRDS LIFE SIZE

1 BLACKTHORN (SLOE)   1A Blossom   2 SHEPHERD'S BULLACE   2A Blossom
3 FARLEIGH DAMSON   4 PRUNE DAMSON
5 GREENGAGE   6 TRANSPARENT GAGE

PLUMS (2): COOKING VARIETIES

1 Cherry Plum (*Prunus cerasifera*). This is a distinct species. It has brightly coloured red, bronze, or yellow fruit, which are round with a small point, shiny without bloom and with a soft juicy flesh rather tasteless when picked but improved with cooking. It ripens in July and August. When not in fruit it is readily distinguished by its shiny leaves, its smooth green shoots, and brown bark. The white flowers open with the leaves. A native of Western Asia, it grows readily in this country and is often used as a hedge or windbreak instead of quickthorn, but it rarely sets a heavy crop. The purple-leaved, ornamental plums with pink or white flowers are varieties of this species. There is also an ornamental fruiting kind 'Trailblazer' with dark foliage, white flowers, and crimson fruit.

2-6 Plums. The numerous European Plums are generally classed under *P. domestica*, which is a hybrid between the forms *P. spinosa* (p. 67) and *P. cerasifera* described above. The two species grow together in Asia Minor and have interbred for thousands of years. Cultivated plums show great variety in their colour, shape, size, and flavour — indeed, the variety that one would expect from the progeny of two such distinct parents. Less obvious but just as interesting is the variety in leaf colour and form, in the twigs, and in the general shape of the trees. Many types of plum breed fairly true from seed. There are consequently families whose members are often difficult to distinguish one from the other. Many of our best plums seem to have originated as wildings in woods and hedgerows, and in most cases their history has not been recorded. The nineteenth-century pomologists Knight, Laxton, and Rivers made great progress in the raising of new varieties by deliberate breeding, a process continued in this century by Spinks at Long Ashton who raised 'Severn Cross' and 'Thames Cross'.

Plums grow readily from their stones, and in plum areas many self-sown seedlings occur which, not being grafted, sucker freely. Suckers of promising sorts were distributed locally, and in villages all over the country one comes across local varieties known only in their small area.

Cooking plums are characterised by their acidity and rather juiceless flesh. Though there is no relationship between colour and quality, the main families of cooking plums have blue-black skins. A smaller number are greenish-yellow or purple-red. They are invariably larger, more vigorous trees than the choicer dessert sorts, hardier in growth and flower, and heavier, more reliable croppers. The fruit is stewed, used in pies and preserves, and canned.

2 'Rivers' Early Prolific' and **'Czar'** are the best of the several round black early sorts. Though relatively small, they bear heavy crops. 'Black Prince' is a newer variety of this type. 'Belgian Purple' and 'Early Orleans' are older kinds now superseded.

3 'Pershore Egg' is well-known as the most prolific, grown on a large scale in the Pershore and Evesham area. It is a greenish-yellow plum of poor dessert quality but excellent for jam. There is also a red-fruited bud sport known as 'Red Pershore'. 'Gisbornes' is an older sort of the same type. 'Warwickshire Drooper' is also similar but of better flavour. 'White Magnum Bonum', also known as 'Yellow Egg Plum', is an older variety, probably a parent of all this family.

4 'Pond's Seedling' is another prolific plum, hard, and travelling well when ripe. 'Blaisdon Red', even harder, is of poorer quality but is a good example of a local disease-resistant plum, and a valued fruit in its own area. 'Giant Prune' is probably a seedling of 'Pond's Seedling', similar in character but juicier. 'Red Magnum Bonum', an old rather dry sort, is a relation, as is 'Autumn Compote', one of the choicest preserving and drying plums.

5 Monarch, an older variety with a fairly large almost round fruit, is grown for late markets, as is the rather similar 'Wyedale', and the newer, popular 'Marjorie's Seedling'.

6 'Prune d'Agen', 'Prince Engelbert' and **'Fellemberg'** are characteristic of the large family of prunes for drying. They are late, oval, black-skinned fruits, mostly needing a hotter climate than Britain's to dry on the tree but easily dried artificially. The Kentish variety 'Diamond', and the Yorkshire 'Winesour' are prune-like in flavour when cooked but ripen earlier. The oldest of this section is perhaps the 'Quetsche', a common plum of Central and Southern Europe, often doing well in England. The 'Mussell Plum' is similar but earlier. Plums thrive best in a continental climate — one of cold winters and hot summers. In the districts where plums are difficult, heavy crops often follow a long hard winter. In Britain they are more successful in those parts least influenced by the Gulf Stream. They are chiefly grown in Kent, Middlesex and the Thames Valley, East Anglia and Cambridgeshire, and the border counties Worcester, Gloucester, Hereford, and Shropshire. Because they grow so readily from stones and crop so freely, they are seen in most gardens in favourable areas and occur widely in hedges and woods. As they naturally tend to droop with age, the most natural form in which to grow them is as a small standard; but in gardens they may be grown as bushes, or sometimes as pyramids, and for the choicer sorts, as fan-shaped trees on sunny walls. They are budded or grafted on to specially grown stocks — standards on Myrobolan, and dwarfer trees on St. Julien A rootstocks.

TWO-THIRDS LIFE SIZE

1 CHERRY PLUM 'MYROBALAN'   1A Blossom

PLUM VARIETIES

2 'RIVER'S EARLY PROLIFIC'   3 'PERSHORE EGG'

4 'POND'S SEEDLING'   5 'MONARCH'   6 'PRUNE D'AGEN'   6A Dried Prune

69

PLUMS (3): DESSERT VARIETIES

A richer flavour and a higher sugar content differentiate dessert plums from the cooking kinds. Where the origin of these sorts is known, it will be found that one parent is a gage and though the gages are the choicer fruit, at least some of these plums come near to them in richness. Many dessert plums are, like the gages, green or golden; a few are black; and a few others red or purple. Most of them, again like the gages, are relatively susceptible to climatic conditions and in climates like Britain's will only give their best fruit in sheltered conditions, in favourable situations, and when trained on a wall. Earliness is a valuable quality for although there are plenty of plums in September there are few when fresh fruit is scarce in August. The earliest variety is 'Early Laxton', a small, bright orange and yellow fruit. It probably comes from a similar older Spanish variety 'Jaune Hative', known in the seventeenth century. 'Ontario', a Canadian variety of gage parentage, and 'Opal', a purple Scandinavian variety, ripen in mid August.

1 **'Victoria'** ripens in late August. This famous red plum, although not of first-class flavour, is important as being the only hardy, self-fertile, heavily cropping dessert variety which will grow under orchard conditions in Britain. It was found in a Sussex wood about 1840. At about the same season are 'Oullins Golden Gage' and 'Goldfinch', which are oval yellow plums hardy as orchard trees and heavy croppers.

2 **'Coe's Golden Drop'.** One of the choicest dessert plums, it is, unfortunately, a light cropper even in the best conditions; it should be planted on a sunny wall. Two good crosses have been raised from it at Long Ashton — 'Thames Cross' which ripens in early September, and 'Severn Cross', for late September. These are both vigorous and self-fertile.

In breeding new plums, raisers have generally tried to cross a choice dessert plum with one which is more prolific though of poorer quality, with the aim of producing a prolific tree with first quality fruit. For this reason Coe's Golden Drop has been much used as a parent. A sport from this plum, the violet-coloured 'Coe's Violet', has also proved itself a successful parent; 'Merton Gem' has been raised from it by the John Innes Institute; this is a round red dessert sort for mid September.

3 **'Kirke's Blue'.** Of the black dessert sorts 'Kirke's Blue' is the choicest, but like 'Coe's' it needs good conditions and will only succeed if planted on a sunny wall. Similar in character are 'Angelina Burdett' and 'Woolston Black', both raised near Southampton by a Mr. Dowling, and 'Count Althann's Gage', a dark crimson-red gage-like fruit. These are all September sorts.

4 **'Laxton's Delicious'** is also a cross from 'Coe's Golden Drop' raised by Laxtons. It ripens in mid September. 'Anna Späth', with the same season, is a newer introduction of Hungarian origin with oval crimson-purple fruit.

5 **'Jefferson'** is unusual in that it is an American sort which has proved hardy and adaptable in England. Raised over 150 years ago, it is one of our choicest fruits, as good as a gage in warm seasons. It has produced a red-fruited bud sport known as 'Allgroves' Superb'.

Japanese Plum. Some of the dessert plums grown in South Africa, the U.S.A., and Australia, are derived from the Japanese Plum (*Prunus triflora*, or *P. salicina*), which in spite of its name is a native of China. It is more tolerant of climatic variations and more profitable than the European plum but although it is brightly coloured orange-red or golden yellow, it is usually of inferior flavour. In the U.S.A. many hybrids with other species have been raised with a view to improving the flavour of this type.

American Plums. Many American plums have been derived in recent years from native species, and the native plums are also sometimes grown. These include the 'Chickasaw' (*P. angustifolia*); *Prunus americana*; and the 'Oregon plum' (*P. subcordata*). These are all red-fruited species. Also native to the States and edible are the 'Texan plum' (*P. orthosepala*), and the 'American sloe' (*P. alleghaniensis*), both of which have black fruits. These native fruits withstand the violent changes of temperature, and endemic diseases, and native insect pests better than do the choicer exotic kinds which can only be grown in fruit areas.

The Black European Prunes such as 'Fellemberg' and 'D'Agen', as well as native seedlings from these varieties, are grown widely in California, Washington, and other Western States, (as they are also in Australia). Eighty per cent of the U.S.A.'s plum trees are grown on the Pacific Coast.

TWO-THIRDS LIFE SIZE

PLUM VARIETIES

1 'VICTORIA' 1A Blossom 2 'COE'S GOLDEN DROP'
3 'KIRKE'S BLUE'
4 'LAXTON'S DELICIOUS' 5 'JEFFERSON'

PEACHES AND APRICOTS

Both the peach and the apricot originated in China. Modern varieties are not very different from those cultivated by Roman and Chinese gardeners two or three thousand years ago. Over the whole warm temperate zone from China through India, Persia, the Levant, and the Mediterranean area, naturalised populations are found which are relics of ancient cultivation.

1-2 Peach (*Prunus persica*) is a small, willowy, deciduous tree, often short-lived. Its flowers (1A) are either small and rose-pink or larger pale pink, rarely white, produced on year-old wood and opening about a fortnight later than the almond (p. 29). The relationship between these two trees is obviously close but the latter has thicker leaves, a leathery instead of juicy fruit which splits along the suture, and a smooth not prickled stone. The two hybridise readily.

Peaches vary considerably in their season of ripening, size, and colour. The velvety skin varies from greenish-white to golden-yellow, with a crimson colouring which may be almost non-existent or a complete covering. The colour of the flesh may be greenish, white or yellow: the white-flesh type is usually considered the best flavoured and the hardiest in a cold climate; the yellow-fleshed kinds are most popular for canning and many have a flavour as good as that of the white fleshed kinds. They mostly need a warm climate. Peaches are grown widely in temperate regions, particularly in the Mediterranean countries, the United States, South Africa, China, Japan, Australia, and New Zealand. They are delectable dessert fruit, but need great care in handling, transporting and storing. They are very popular canned, dried, and made into jam. Some are also now frozen. Most of the peaches eaten in the British Isles are imported, but some are grown under glass and even out-of-doors, especially on walls facing south and west, in areas sheltered from rain, summer gales, and spring frost. The sorts usually grown are the early white-fleshed 'Duke of York' which was raised by Rivers in the 19th century; for midseason, either 'Peregrine' (1) another of Rivers' raising which has yellowish-white flesh, or the yellow-fleshed 'Rochester' (2) a newer variety from the Rochester Experimental Station in New York State; and for a late variety the 'Royal George' which has yellowish-white flesh.

3 Nectarine (var. *nectarina*). This is a smooth-skinned peach of richer flavour, brighter colour, and generally smaller size. Stones sown from one tree may grow up to be the other — and nectarine fruits are even said to occur as bud sports on peach tree branches. Three white-fleshed varieties, all raised by Thomas Rivers, are: 'Early Rivers', an early kind; 'Lord Napier' (3) for midseason; and 'Pineapple', a late variety.

4 Apricot (*Prunus armeniaca*). This is a sturdier tree, 20 to 30 feet high, with white and only rarely pink flowers borne in March before the leaves appear on two-year-old and older wood. Its fruit varies in the depth of its colouring, from pale yellow to deep orange with a red freckled skin, and also varies in size, flavour, and tenderness. In suitable conditions apricot trees grow readily from discarded stones. They need a warm temperate climate, and must be guarded from frost. They are grown mainly in China, Japan, North Africa, and California, and have the same uses as peaches. When apricots are grown in England, the variety is usually either 'Farmingdale', a newer Canadian sort, or the old English variety 'Moorpark' (4).

Cultivation. Like other deciduous trees, peaches and apricots need a dormant period, varying in length with the variety, and their culture is only possible in countries with a cool winter. Though they may all be grown on their own roots or on seedling peach and apricot stocks, they are most successful in Britain if budded on the plum stock St. Julien A. There is no means known of dwarfing these trees, except by growing them in pots, and as they will grow strongly it is important to plant them only where there is space for them to develop properly. Peaches flower and bear fruit chiefly on the shoots formed in the previous year, so it is essential to keep the tree growing strongly if it is to bear successful crops. Old peaches which have become unfruitful are often sawn off to obtain new strong shoots from the trunk.

TWO-THIRDS LIFE SIZE

1 PEACH 'PEREGRINE'   1A Blossom   2 PEACH 'ROCHESTER'
3 NECTARINE 'LORD NAPIER'
4 APRICOT 'MOORPARK'   4A Blossom

STRAWBERRIES

These greatly valued fruits of temperate climates are delicious but perishable. Modern cultivated strawberries all derive from American species. Mainly a dessert fruit, they are also canned, frozen, and made into jam.

1 Strawberry (*Fragaria × Ananassa*, etc.). The juicy edible part of the strawberry fruit is an enlarged receptacle on the surface of which the achenes or seeds are embedded (1c). The strawberry plant is a perennial herb, with a leafy crown from which radiate prostrate stems or runners bearing small leaf clusters which take root and will grow into new plants.

2 Alpine Strawberry (*F. vesca semperflorens*) is a variety of one of the European wild strawberries (*F. vesca*) which is widely distributed in woods and shady grasslands of the northern hemisphere. *F. vesca* has small, richly flavoured fruit, less acid than the usual garden strawberries, in early July; the alpine strawberry has rather larger fruit which ripens over the summer months. 'Baron Solemacher' is a good runnerless or bush form, with typical small red fruits (one variety of Alpine strawberry has white fruits). *F. vesca*, although cultivated for centuries, did not greatly improve in the size or flavour of its fruits, any more than did another more rarely cultivated European wild species, the Hautboy (*F. moschata*). The latter makes fewer runners and has rounder, pinkish-red musk-scented fruit on which the achenes are set more densely towards the end of the fruit. Both of these may be raised from seed, and birds do not take their fruit.

3-5 Cultivated Strawberry Varieties. The cultivated strawberry as we know it is of American origin. The first species to be introduced into Europe soon after 1600 was *F. virginiana*, the scarlet woodland strawberry of the Eastern States. It spreads quickly to make a thick carpet of leaves and may sometimes be seen covering large areas of railway embankment. Under the name of 'Little Scarlet' it is still grown for jam making, its firm, sweet, red-fleshed fruit being particularly good for this purpose.

More than a century later the West Coast Pine strawberry (*F. chiloensis*) was introduced. Though first seen in Chile, it grows in several countries along the Pacific seaboard of North and South America, often in sand on the beach. It has thick dark foliage and pinkish or white fruit with a marked pineapple flavour. Male and female flowers are borne on separate plants, and female plants will not set fruit unless planted with pollinators.

Hitherto divided by mountain ranges, these two species interbred freely when grown together in Europe and from 1800 onwards new hybrids which combined the characteristics of both parents were on sale. In 1892 Thomas Laxton bred the famous variety 'Royal Sovereign' (3), a sort which has remained a standard of excellence to this day. Resistance to disease and the nature of virus infection was but little understood by earlier breeders and most of their splendid varieties have now died out. Almost alone 'Royal Sovereign' has been retained and this only as the result of much scientific skill in raising stocks free of virus.

In strawberry-growing countries new sorts are continually being raised by breeders in the search for more vigorous plants adapted to local climates and better crops. In England the most notable new sorts of, recent years are the Cambridge varieties. More than half of this country's acreage is now planted with 'Cambridge Favourite' (4); and others such as 'Cambridge Vigour' (5), 'Rival', 'Late Pine', and 'Rearguard' are widely grown. From the Scottish Clydeside Research Institute have come 'Auchincruive Talisman', 'Templar', and 'Red Gauntlet'.

The perpetuals or remontants are an interesting group, which flower successively during the summer and produce fruit from July till October. Best known are 'La Sans Rivale' and the newer 'Hampshire Maid', which grow few runners or none at all. There is, however, also a runner-bearing remontant known as the 'Climbing Strawberry'.

Varieties differ in their ability to bear a second crop. Some, such as 'Red Gauntlet' and others of American origin, form a second crop of flower during the hot months and consequently ripen fruit in August and September; others do not seem able to form flowers until it is too late for their crop to ripen. Cutting back or burning over after fruiting helps development of a second crop and at the same time destroys insect pests.

Culture. In the past commercial culture was concentrated around cities. With the coming of railways, warm sheltered areas such as the Tamar and Cheddar Valleys where early crops could be raised out of doors became important, but the increase of air transport and the invention of cheaper plastic cloches has somewhat reduced their importance.

Virus diseases are a great danger to strawberries. Spread by aphis, they rapidly infect and ruin healthy fields. For growers' stocks, much care and skill is now devoted to the maintenance of virus-free strawberries raised in insect-proof houses or in mountain areas where aphis does not exist.

As a garden crop strawberries are most rewarding, bearing a good crop in their first season. Almost any soil can be enriched and improved to please them. As for other surface-rooting herbs, good cultivation and ample moisture are essentials. Many ingenious methods of cultivation have been devised. They may be grown in barrels, on artificial banks, protected by mats and sheets — but most people will grow them in the traditional way, on the flat with straw set under their flower stems as the fruit sets and well protected with net to keep off birds.

1A

1B

2

1

1C

1

3

4

5

PLANT × ¼   *FRUITS AND FLOWERS LIFE SIZE*   *FLOWER SECTION* × 2
1 STRAWBERRY PLANT   1A Flowers   1B Flower section   1C Section of fruit
2 ALPINE STRAWBERRY 'BARON SOLEMACHER'
STRAWBERRY VARIETIES
3 'ROYAL SOVEREIGN'   4 'CAMBRIDGE FAVOURITE'   5 'CAMBRIDGE VIGOUR'

RASPBERRIES

Raspberries, blackberries (p. 79) and other related berries are all species of the genus Rubus. The fruit or berry is one composed of numerous round one-seeded drupelets which are closely set together on a small conical receptacle or core. The manner in which these drupelets cohere and part from the receptacle is most noticeable when the fruit is being picked.

Raspberries are delicious fresh or cooked, and as a basis for jams, jellies, and drinks. Large quantities are canned and frozen.

1-3 Raspberry (*Rubus idaeus*) has a shallow widely spreading woody root system. From adventitious buds, upright suckers or canes grow forming a thicket of stems. These canes bear flowers either in their first autumn or in their second summer. After fruiting the cane dies and the whole cane should be removed as soon as fruit has been gathered.

This hardy species grows wild in Britain and in much of the Northern hemisphere, most often in high hilly and heath country in acid soil. From the flowers a richly scented honey is gathered. Many colonies seen growing wild may have escaped from cultivation, for raspberry seeds are widely spread by birds. In fruit farms, self-sown seedlings are troublesome. The most frequent type bears red fruit in July on second year canes, but there are also distinct varieties with yellow fruit often known as white raspberries, and autumn-fruiting varieties which bear fruit in September and October on the current year's canes.

Of the red summer-fruiting varieties, the oldest named sorts such as 'Antwerp' have now disappeared and even sorts widely grown during the first part of this century have mostly so deteriorated through virus infection as to be worthless. 'Norfolk Giant', a vigorous rather late-ripening sort, has been re-introduced from New Zealand where virus-free stocks were discovered during the Second World War. Most of the other sorts now planted were raised at East Malling Research Station and introduced after 1945. For heavy crops, 'Malling Exploit' and 'Malling Promise' (1) are most widely grown. 'Malling Jewel' is less vigorous but has a better flavour. 'Lloyd George', which was discovered in a wood near Corfe Castle, has two seasons of fruiting: some of its canes bear fruit in their first autumn, and both these and the remainder will crop the following summer. Of all varieties 'Lloyd George' has the best flavour.

The best autumn-fruiting sort is the American 'September' (2). An older sort, 'Hailsham', is also sometimes seen. In these sorts most of the canes bear fruit in September and October; and if they are not removed after fruiting, they may also bear some fruit the following summer. In some dry soils these are the most satisfactory sorts to grow because they ripen at a time when moisture is more plentiful.

The yellow-fruited raspberries (3) are a distinct strain of great antiquity, easily recognised when not in fruit by their pale green leaves and their less vigorous lightly coloured canes. Their fruit is peculiarly soft and richly flavoured. Of several sorts 'Golden Everest' (3) and 'Brynes Apricot' are the best. Though raspberries are grown commercially in many parts of the British Isles the largest farms are in Scotland, particularly in the Blairgowrie area where the cool climate and the light neutral soil seem to be particularly suitable. Most garden soils suit them well though they may be difficult in heavy and alkaline soils.

In the United States, *R. strigosa*, a species with both red-fruited and yellow-fruited varieties, closely related to *R. idaeus*, is often used as a source for cultivated varieties. A black-fruited species *R. occidentalis* is also widely cultivated. This is clump-forming rather than spreading, its canes are heavier and more branching, and the stems, leaves, and fruit have a glaucous bloom on them.

At the East Malling Research Station hybrids between the red- and the black-fruited species have been raised and are known as Black-Red Raspberries. Their fruit is dark purple, of good flavour, and with small seeds which makes it a good jam fruit. Like *R. occidentalis*, it forms clumps of a few large branching canes.

4 Wineberry (*R. phoenicolasius*) is another species of the same section of the genus. It comes from North China and Japan. It is a clump-forming plant, with long arching canes clothed in red hairs, without prickles. Its pleasantly flavoured berries are golden or orange, shiny and translucent, ripening in early August. This is a beautiful species which can be grown on a trellis or fence with little trouble.

LIFE SIZE
1 RASPBERRY 'MALLING PROMISE'   1A Flower section
2 RASPBERRY 'SEPTEMBER'   2A Flowers
3 RASPBERRY 'GOLDEN EVEREST'   3A Flowers   4 WINEBERRY   4A Flowers   77

BRAMBLES AND RELATED BERRIES

Of the innumerable species and hybrids of blackberries found growing wild in the Northern hemisphere, only a few bear fruit sufficiently good to justify cultivation.

1-2 Blackberry (*Rubus ulmifolius*). Of the blackberries native to Britain, this is one of the most worth picking. It is distinct in not being self-fertile and will only set good crops when cross-pollinated. It is a prickly climbing plant thoroughly at home in the traditional field hedge but now less plentiful than it was formerly. It forms a cluster of radiating canes which, though erect at first, curve and arch downwards when opportunity occurs until they touch the ground. Here the shoot takes root and a clump of new shoots is soon formed. In this way a single seedling may soon spread over a large area. The fruit is borne in large clusters at the ends of the older shoots which after two or three years' cropping die.

The older cultivated blackberries were probably found growing wild and their origin is often doubtful. One of the longest known is the 'Norwood', (2) from the wild species Cut-leaved or Parsley-leaved Blackberry (*R. laciniatus*). 'Norwood' is of relatively small growth, often fruiting well on first year canes, and of excellent flavour. There is now a thornless sport known as 'Oregon' which is even more attractive.

The 'Himalaya Berry' may be American or European. It is certainly the most vigorous, often making canes twenty feet long which live for several years and bear up to 50 fruits on a bunch.

Of modern sorts 'Bedford Giant' and 'Merton Early' fruit early, 'John Innes' is late, and 'Merton Thornless' like 'Oregon' is thornless and will fruit over a long period of August and September. Though once much grown in England, the American blackberries, which derive from *Rubus alleghaniensis* and other more Southern species, grow less readily in the English climate and are often killed by frost in a hard winter. Because large quantities of native blackberries have been available in Britain, there has been little inducement for growers to plant large acreages, but both loganberries and blackberries are grown for jam making and canning in the Pershore area and in other market-gardening areas. In the garden they may be easily grown on poles or wire fences where their canes may be tied out in an orderly fashion allowing light and pickers to reach the fruit readily. Many stocks have deteriorated because of virus infection, and it is important to plant clean stock.

3 Loganberry (*Rubus loganobaccus*). The dull red fruits are more acid than blackberries. Though the loganberry is similar to the blackberry in its growth, its leaves are larger and softer and its shoots trail on the ground unless tied up. For many years its origin has been in doubt. It originated in California, but the earliest information that it was a hybrid between a blackberry and a raspberry was doubted on the grounds that it closely resembles another American species, *R. vitifolia*; recent experiments in breeding, however, have shown that it is probably a true hybrid. In the U.S.A. it is grown on a commercial scale for canning. The canes are heavy croppers and may continue to bear for about 15 years. The virus-free Loganberry, LY 59, from Long Ashton is probably the best of the thornless types for producing good crops. There are thornless variants which are easier to handle.

Boysenberry, Veitchberry, and **Phenomenal Berry** are all similar to the loganberry both in their parentage and in appearance.

4 Dewberry (*R. caesius*). This is a more slender plant than the blackberry with smaller, less coherent fruits which have a white bloom and are not shiny. Its trailing stems lie on the ground. The crop is usually light but may nevertheless be valuable since it ripens ahead of the blackberries. The dewberries grown in America have larger fruit than those grown in England. The cultivated forms have been derived from several American species including *Rubus alleghaniensis*.

5 Cloudberry (*R. chamaemorus*), is included because its fruit is so often mentioned in the literature of the Northern countries. It is a little herb which inhabits large areas of open moorland in Canada, Northern Britain, the Outer Hebrides, Northern Europe, and Arctic Russia, often fruiting within the Arctic Circle. Its small golden fruit may be used, as blackberries are, in puddings and for jam.

LIFE SIZE

1 WILD BLACKBERRY   1A Sections of flower and fruit
2 BLACKBERRY 'NORWOOD'   2A Flowers   3 LOGANBERRY
4 DEWBERRY   5 CLOUDBERRY

79

CURRANTS AND GOOSEBERRIES

These are both species of Ribes which is the only genus in the order Grossulariaceae. Their fruit is the familiar round berry, with a thin, often translucent, skin enclosing a number of seeds in a juicy flesh. They are shrubs, with three- or five-foliate leaves, bearing their small greenish flowers from buds on shoots of the previous season's growth.

1 Black Currant (*R. nigrum*) grows wild across the whole of Europe and Northern Asia. It is easily recognised at any season by the peculiar heavy aroma given off by its stems and leaves. The older kinds were probably found wild — and a currant collected in Central Asia earlier in the century seemed identical with the usual commercial sorts. In recent decades private firms and research stations have raised thousands of seedlings with a view to raising better sorts. From East Malling we have 'Amos Black' and 'Wellington XXX'; from Long Ashton, 'Mendip' (1), 'Tor', and 'Cotswold Cross'; while Laxton Bros. introduced 'Laxton's Giant'.

The black currant succeeds best if it has several stems and the opportunity to renew its shoots from basal buds. It must therefore be grown in the open as a multi-stemmed bush. It is essential to start with virus-free stock. Since the vitamin content of black currants was first appreciated, the acreage devoted to this fruit has been greatly increased and large black currant farms planted.

2 Red Currant (*R. sativum*). The cultivated red and white currants are nearly all derived from a wild type of the species *R. sativum*. This parent is a plant of Western Europe, chiefly found by streams and in wet woodlands. It has hanging flower stems. The old 'Dutch' currant and 'Laxton's No. 1' (2) belong to this type.

Some varieties, including 'Rivers' Late Red' and 'Red Lake' have as a parent a second species, *R. petraeum*, which has its origins in Central and Southern Europe. This species is easily recognised by its purple-tinted flowers.

At least one old variety, 'Raby Castle', is descended from a third species, known as the Northern Red Currant or British Red Currant (*R. rubrum*). This species is found wild over the greater part of Eurasia in rocky places and mountain woods, and is much cultivated in Scandinavia and Russia where its hardiness is important.

Unlike the many-stemmed black currant, the red currant has a stout main stem and may therefore be grown in standard or cordon form, or fan-trained against a wall. It will continue to grow on a single stem for many years.

3 White Currants are seedlings from the red lacking the red pigment. Though they are so closely related, they have a distinct flavour, being rather less acid than the red kinds.

4-6 Gooseberries (*R. grossularia*). Those grown in Europe (including Britain) derive from a plant native to most of Europe, especially in shady valleys and rocky woods. It varies much in habit of growth — from widely spreading or pendulous, to almost upright — and in the size, colour, skin, flavour, and shape of the fruit. A yellowish type with a hairy skin seems to predominate, but varieties with reddish, dark green, and pale whitish-green skins are also seen. Some of the best flavoured sorts have a downy dark green skin even when ripe. This wide variety has given rise to a large number of named sorts most of which were introduced during the nineteenth century; the most famous of these, 'London', had the unique distinction of providing the largest exhibited fruit for 37 unbroken seasons from 1829 to 1867. For cooking purposes, 'Careless' (4), 'Whinham's Industry' (5), Howard's 'Lancer', 'Leveller', and 'Whitesmith' are perhaps best. For dessert gooseberries 'Early Sulphur' (6), 'Warrington', 'Langley Gage', 'Glencarse Muscat', and 'Whitesmith' are choicer.

Gooseberries are chiefly grown on fruit farms near towns, as a second crop among orchard trees, and in market garden areas especially near jam factories. They are easily grown in gardens, but repay high culture.

The most serious problem facing gooseberry growers is the American gooseberry mildew, a blighting disease which ruins both gooseberries and black currants unless preventive measures are taken. But it only affects European species, for American species have presumably built up a resistance to it.

For this reason Americans grow hybrids raised from native species.

7 American Gooseberries. One American species is *R. divaricatum*, the so-called 'Worcesterberry' because it seems first to have been sold by a nurseryman in Worcester who thought it was a black currant gooseberry hybrid. It is a small black gooseberry and immune to mildew. While the true hybrid is worthless, this is a valuable fruit. The strong young canes, which may be four feet long, are arched, and up to twenty berries hang singly from their underside.

Another important American native, *R. hirtellum*, has also been known as the 'currant gooseberry'. It also has red-purple fruit like small gooseberries.

'Pixwell' (7) is the best known of the hybrid American gooseberries. It has long arching canes, and light green fruit hanging on long stalks well below the thorns and foliage. The American types are all smaller than British gooseberries, but they may please anyone who wishes to avoid spraying.

LIFE SIZE   FLOWER SECTIONS × 3

1 BLACK CURRANT 'MENDIP'   1A Flowers   1B Flower section
2 RED CURRANT 'LAXTON'S No. 1'   3 WHITE CURRANT
4 GOOSEBERRY 'CARELESS'   4A Flowers   4B Flower section
5 GOOSEBERRY 'WHINHAM'S INDUSTRY'   6 GOOSEBERRY 'EARLY SULPHUR'
7 GOOSEBERRY 'PIXWELL'

81

FRUITING SPECIES OF THE ERICACEAE

1 Bilberry (*Vaccinium myrtillus*). The fruits of this native of the British Isles and many parts of Europe and northern Asia are variously known as bilberries, blaeberries or whortleberries. They are acid when raw, but palatable in tarts or when made into jam. They tend to be rather neglected, because picking them is a laborious and time-consuming task, in comparison with other wild fruits such as blackberries.

Bilberry is a low shrub, up to 2 feet high, growing abundantly on heaths and moors and in acid, open woodlands. It has green, angled twigs, bearing ovate, acute, serrulate leaves, about $\frac{1}{2}$ to 1 inch long. The greenish-pink to pink, globose flowers are borne singly or in pairs in the axils of the leaves. The juicy, bluish-black fruits are globose and about $\frac{1}{4}$ inch across, ripening from July to September.

2 Highbush Blueberry (*Vaccinium corymbosum*) is a native of North America, where blueberry pie is a traditional dessert and the fruits are also eaten stewed with sugar or in steamed puddings.

The Highbush Blueberry thrives best in moist, acid, peaty soils. It is sometimes grown in Britain. In its native habitat it forms a bushy shrub, up to 15 feet high, with ovate, pointed leaves, 1 or 2 inches long. The white or pinkish flowers (2A, 2B), up to $\frac{1}{2}$ inch long, are borne in short racemes. The bluish-black fruits are about $\frac{1}{3}$ inch across on wild plants, larger on selected cultivars, many of which are probably of hybrid origin, derived from this species and from the Rabbit-eye Blueberry (*V. ashei*) and the South-eastern Highbush Blueberry (*V. australe*).

Lowbush Blueberry (*V. angustifolium*) is another North American blueberry which is grown commercially in the north-eastern States and Canada. Being more northerly in distribution, it is hardier than the high-bush blueberries. It is a low shrub, about 1 foot high, with finely-toothed, lanceolate leaves, $\frac{1}{4}$ – 1 inch long. The corolla is greenish-white, sometimes with reddish streaks. The bluish-black fruits are $\frac{1}{4}$ – $\frac{1}{2}$ inch across.

3 Cranberry (*Vaccinium oxycoccus*) is a native of Europe (including the British Isles), North Asia, and North America. Its acid fruits are used mainly for making cranberry sauce, served with turkey or venison.

The cranberry is an evergreen, prostrate undershrub with wiry stems, bearing oblong-ovate leaves, less than $\frac{1}{2}$ inch long, dark green above, glaucous beneath, with revolute margins. Its terminal pink flowers (3A) have 4 strongly recurved petals. The bright red, rounded or broadly oval fruit is about $\frac{1}{3}$ inch across.

4 Large or **American Cranberry** (*Vaccinium macrocarpon* Ait.). This American species is sometimes cultivated and occasionally naturalized in the British Isles. Its uses are the same as those of the previous species, which it resembles in habit although it is more robust, with larger, less revolute leaves, flowers in lateral clusters, and larger fruits, up to $\frac{3}{4}$ inch across.

Cowberry (*Vaccinium vitis-idaea*). Also known as the Mountain Cranberry, this shrub with edible berries is native in the British Isles and throughout most of the cooler parts of the northern hemisphere. It is an evergreen shrub, 6 to 12 inches high, with dark green, glossy, obovate leaves, $\frac{1}{5}$ to $1\frac{1}{5}$ inch long. Its white or pinkish, campanulate, 4-lobed flowers are borne in small terminal racemes. The red globose fruit is about $\frac{1}{3}$ inch across.

5 Arbutus or **Strawberry Tree** (*Arbutus unedo*). This is a native of the Mediterranean region, and also locally in Ireland. It is grown for ornament in British gardens. The fruit looks appetising but is disappointingly unpalatable, although harmless. In France, it is used in making liqueurs and confectionery.

Arbutus unedo is a shrub or small tree, up to about 30 feet high. It is evergreen, ovate to obovate, serrate leaves, 2 to 4 inches long. The creamy-white or pinkish flowers (5B), about $\frac{1}{4}$ inch long, are borne in broad, drooping panicles (5A), from October to December. The fruit, a red, warty berry, about $\frac{3}{4}$ inch across, matures a year later.

LIFE SIZE   *FLOWER SECTIONS* × 3

1 BILBERRY   1A Immature fruits   2 BLUEBERRY   2A Flowers   2B Section of flower
3 CRANBERRY   3A Flowers   3B Section of flower   4 LARGE CRANBERRY
5 ARBUTUS   5A Flowers   5B Section of flower

The citrus fruits are one of the largest and most important groups of fruits of the tropical and sub-tropical regions. The group as a whole originated in China and south-east Asia, though the lime was perhaps native to India, and some hybrid fruits have been developed in more modern times in various parts of the world.

The members of the genus *Citrus* are all small trees or shrubs which can be recognised by certain botanical characteristics such as the more or less winged leaf-stalk and the presence in many species of a small spine in the axil of the leaf. Three plates are devoted to the species of citrus, which range from popular fruits which are produced on an enormous scale such as the orange and grapefruit to others which are of only local importance such as the citron and kumquat. The attraction of the fruits for eating depends on the fact that their juice contains, on a dry matter basis, 80 – 90 per cent of sugars and acids. Sugars predominate in most species, making them very pleasant to eat as fresh fruit, but in the lemon and lime acids predominate so that they are used in rather different ways. Another valuable nutritional point is the richness of all citrus fruits in vitamin C, although ironically enough lime-juice, which was so often used on ships to prevent scurvy, has a lower content of this vitamin than do oranges, lemons, and grapefruit. All the species contain essential oils in the skin of the fruit and often in the flowers and leaves, which are sometimes extracted for use in perfumery and soap-making; the bergamot (*Citrus bergamia*) is in fact mainly grown for this purpose. Other products which are sometimes obtained from the fruits include citric acid (now more often produced by other processes) and pectin, which is used in jam-making. One species, *Citrus trifoliata*, because of its thorny nature, is sometimes used as a hedge plant. Although all citrus species can be grown from seed, they are very commonly propagated by budding to ensure that desirable fruit characters are reproduced in the new plant without genetic variation.

1-2 Sweet Orange (*Citrus sinensis*) is a tree growing up to about 25 feet high. It was introduced into Europe during the first millennium A.D. It is nowadays grown on the largest scale in warm regions outside the tropics, especially in those with a Mediterranean type of climate. As with all citrus fruits, frost is a hazard though the orange is more resistant to cold than lemons, grapefruit, and limes. In the tropics, the fruits often fail to develop the typical 'orange' colour and ripen with the skin still green, which makes them less attractive on the export market though the orange colour can be developed by suitable gas treatment. Among the best-known areas producing oranges for the world's markets are Spain, Israel, Florida, California, and South Africa. There are also considerable acreages in Australia, Brazil, and the North African countries, while some oranges are grown for domestic consumption in almost every country with a warm climate. Many of the favourite varieties of orange are very well known on world markets. These include; 'Jaffa', whose fruits are large, juicy and seedless; 'Valencia', a late variety with a long fruiting season; and 'Washington Navel', whose large fruits have a navel-like mark at the apex and are almost seedless. In the 'blood oranges' (2) the flesh has a blood-red tint; well-known varieties include 'St. Michael' and 'Maltese'. Outside the tropics, most oranges come to maturity in the winter, the main harvesting period in the northern hemisphere being from about October to March. Sweet oranges, with their sugary pulp, are mostly consumed as fresh fruit or used in preparing orange juice drinks. Orange flowers (1A) have a delicious scent and are sometimes used in decorations and bouquets.

3 **Seville, Bitter, or Sour Orange** (*Citrus aurantium*) bears fruits with a bitter taste which makes them unattractive for eating raw. The fruit is largely used in manufacturing marmalade, an orange conserve which finds its greatest popularity in Britain. There is a large commercial production in Spain, to which the fruit was introduced from Asia by the Arabs, and British imports of sour oranges come mainly from that country. In the past, the sour orange has been much used as a hardy rootstock on to which to graft other species of citrus, but it has been less popular for this purpose since it has been discovered to be susceptible to the 'tristeza' disease.

4 **Lemon** (*Citrus limon*) has a fruit of very distinctive appearance with a blunt excrescence at the apex and a yellower skin than the oranges. It also differs from the oranges in having less sugar and more acid (mainly citric acid) in the pulp. The lemon is very widely grown in Mediterranean and sub-tropical countries, but only rarely in the tropics where the lime replaces it. Italy, Spain, and California are major areas of production. Well-known varieties include 'Eureka', 'Lisbon', and 'Villafranca'. The lemon's acid taste makes it unpleasant to eat as whole fruit but is a culinary virtue; the juice is squeezed over other foods to flavour them, and used in cookery, baking, and confectionery, and in the preparation, with added sugar, of lemon juice as a refreshing drink (lemonade). A variety known as 'rough lemon' is very widely used as a rootstock on which to graft other citrus fruits.

1A

1B

1

2

3

4A

4

TWO-THIRDS LIFE SIZE SECTION LIFE SIZE

1 SWEET ORANGE   1A Flowers   1B Section of flower
2 BLOOD ORANGE   3 SEVILLE ORANGE
4 LEMON   4A Flowers

85

CITRUS FRUITS (2)

1 **Grapefruit** (*Citrus paradisi*). The tree is somewhat larger than the orange tree, growing from 20 to 40 feet high. The fruits are also larger with a yellow skin, and are borne in clusters — from which fact a somewhat fanciful resemblance to a cluster of grapes is said to have given the tree its name. It is the most important type of citrus to have originated outside Asia, and is thought to have arisen in the West Indies either as a sport from the pomelo, a coarser citrus fruit, or as a hybrid between the pomelo and sweet orange. It had been recognized as a distinct species by 1830, and commercial cultivation began in Florida about 1880, but the fruit did not attain world-wide popularity until the twentieth century. The grapefruit and orange are now the two citrus fruits which are marketed in the largest quantities. The grapefruit grows equally well in the tropics and the warm sub-tropics, provided that frost can be avoided. Amongst the most important producing areas are Florida, California, the West Indies, Central America, and South Africa. Most kinds of grapefruit have a pale yellowish pulp, and the most commonly grown variety, 'Marsh's Seedless', is of this type; but there are also kinds with a pinkish pulp, and most of the grapefruit grown in Texas are of this type. Grapefruit are much consumed as fresh fruit, and are so large that a half fruit is a usual portion for one person; sugar is often added, as the pulp is not so sweet as in the orange. There is also a large trade in canned grapefruit segments, which can be used in fruit salads and other dishes. Grapefruit juice is prepared as a bottled drink, and a marmalade can be made from the fruit. For the fresh fruit market, seedless types are commonly grown, and the seeded kinds of grapefruit are largely used in processed form.

2 **Lime** (*Citrus aurantifolia*). This tree is one of the smallest of the citrus species, rarely reaching more than 12 feet in height. Being usually much branched, it has more the appearance of a shrub than a tree. The fruits also are smaller than most kinds of citrus (commonly about $1\frac{1}{2}$ inches in diameter) and have a yellowish-green skin. Because it is one of the least hardy of the citrus species, the lime is mainly cultivated in the tropics and is only found in a few sub-tropical areas where there is no risk of frost. As one of the acid citrus fruits, it plays the same part in tropical diets as does the lemon in sub-tropical and temperate lands. It is largely used for squeezing the juice over other foods to bring out their flavour, or sometimes as an additive in drinks such as the rum punch of the West Indies. Being such a small tree, room can easily be found for it in gardens, and the tropical housewife often has her supply on the spot. Limes are not very much exported, since such small fruit dry out quickly, lose their juiciness, and become wrinkled and un-

attractive. Lime juice, however, is prepared on a large scale as a bottled drink with added sugar; it has very refreshing qualities and is marketed in many countries of the world. The island of Dominica in the West Indies is a major source of limes for this purpose. Among the varieties grown commercially are 'Tahiti' and 'Mexican', but limes are more often planted from seed than most citrus fruits, and many of the trees grown in the tropics are of uncertain genetic origin.

(*Note*: The lime which is a citrus fruit should not be confused with the lime or linden tree (*Tilia spp.*) of temperate lands, which although bearing the same English name is a totally unrelated plant.)

3 **Tangerine** (*Citrus reticulata*). The fruit is also known under several other names, one of which is 'mandarin', denoting its origin in the Far East, where it first became known to Europeans in the eighteenth century. Some authorities have attempted to discriminate between tangerines and mandarins as different varieties, the former having a deeper-coloured skin than the latter, but the nomenclature has become so confused that they are practically used as synonyms. The same applies to the terms 'king orange' and 'satsuma' which were originally applied to particular varieties of tangerine but tend to be used in some areas of the whole group. In southern Africa the Afrikaans name 'naartje' is often used. Tangerine trees tend to be a little smaller than orange trees and have somewhat more resistance to cold, so that they may sometimes be found planted a little further north in such countries as China, Japan, and Spain. The chief areas of commercial production are the south-eastern United States and southern Europe. Typical tangerine fruits are smaller than oranges, with a skin of a deeper orange or reddish colour, which is conveniently loose and easy to detach from the fruit. The pulp is very sweet and juicy and has a characteristic and attractive flavour. Tangerines are mostly eaten as a fresh dessert fruit, and as the main harvesting season includes the Christmas period they are very popular at that time. Tangerine segments are also canned for marketing and export. The so-called 'Rangpur lime', although an acid fruit, is probably more closely related to the tangerines than to the limes, and is often used as a rootstock for budding other kinds of citrus. The 'Calamondin' is another variety of tangerine sometimes valuable as a rootstock because of its resistance to certain diseases. There are a very large number of named varieties of tangerine, but amongst so many none are of outstanding individual importance. This species has also given rise to many hybrids with other citrus species, some of which are mentioned on page 88.

TWO-THIRDS LIFE SIZE
1 GRAPEFRUIT   2 LIME
3 TANGERINE

1 Citron (*Citrus medica*) was one of the first of the citrus fruits to come from the Far East to the Mediterranean area, where it appears to have been introduced about 300 B.C. This is now the main centre of the comparatively small commercial cultivation which is carried on, Sicily, Corsica, and Greece being among the producing areas. The fruit can, however, also be grown in the tropics. The citron is a rather small citrus tree, 6 – 10 feet high. The fruits are of an unusual shape for citrus, elongated rather than globular and sometimes rather resembling the shape of pear or lemon though they can be up to 8 inches long; the skin is often rough and is greenish-yellow to golden-yellow. The fruits have a very thick white inner skin enclosing a small sour pulp, and are mainly used in the preparation of candied peel. This is produced by soaking in brine and preserving in sugar. A peculiar variety is the 'Fingered Citron' with irregular outgrowths on the surface of the fruit. The 'Etrog' type of citron has long been used in the Jewish religion for ceremonial purposes at the feast of the Tabernacles.

2 Kumquat (*Fortunella spp.*) is not a true Citrus but comprises species of the closely related shrubby genus *Fortunella*. In appearance the fruits are like very small oranges, but they have an acid taste. They are not widely grown, but are prized in China because of their cold resistance which enables them to be grown farther north than any of the true citrus fruits. They are used mainly for pickling or making into conserves.

3 Clementine is regarded by some authorities as a variety of tangerine, and by others as a hybrid, probably between the tangerine and sweet orange. It lies in any case somewhere between the tangerine and orange in the size and colour of the fruit and in the ease with which the peel may be removed. Clementines are produced principally in the North African countries and appear on the market in limited quantities.

4 Ugli. This is a hybrid between the grapefruit and the tangerine. Its general appearance is that of a small grapefruit, and it is used for the same purposes as that fruit. The ever increasing popularity of the grapefruit has made it possible for the ugli as a similar fruit to find a market, though as yet only in comparatively small quantities.

Other Citrus Fruits. Although of minor importance, there are a number of other kinds of citrus fruit which are grown and eaten in various parts of the world besides those which we have had room to figure.

The Ortanique is a variety or hybrid of the orange with characteristically flattened fruits, which has attracted attention in recent years because of its good dessert quality. It is grown mostly in the West Indies and is exported from Jamaica.

The Pomelo (sometimes spelt pummelo) or Shaddock (*Citrus grandis*), from which the grapefruit is descended, has the largest fruits of any citrus species. It has a thick skin and a usually rather bitter fibrous pulp. The fruit, however, is rather variable, and trees of the better types are planted on a small scale for local consumption in south-east Asia and occasionally elsewhere in the tropics.

The Papeda (*Citrus hystrix*) is another fruit of the same region which again has a thick skin and a pulp too sour in taste to have become widely popular.

Besides these species, there are very numerous citrus hybrids, many of which have been artificially created by breeders in the United States and grown experimentally without attaining commercial significance.

Tangelos are hybrids between the tangerine and grapefruit, of which the ugli, already described, is an example.

Citranges are hybrids between citron and orange. Tangors are hybrids between the tangerine and sweet orange; one of these, called 'Temple' and marketed as an orange, is grown on a large scale in the United States.

Limequats, Orangequats, and Citrangequats are some of the hybrids between the kumquat and various other citrus fruits. The purpose of hybridization with the kumquat is to produce varieties with improved cold resistance and immunity from citrus cancer.

TWO-THIRDS LIFE SIZE

1 CITRON   1A Flowers   2 KUMQUAT
3 CLEMENTINE
4 UGLI

DESSERT GRAPES

Grapevines belong to the genus *Vitis*. Although this genus includes several species which produce more or less palatable fruits, nevertheless the most important, by far, as a source of dessert grapes, wine grapes, and grapes which are dried as currants and raisins, is *Vitis vinifera* and its numerous varieties or cultivars.

The classification of these numerous cultivars is very complex. They may be grouped as wine grapes as distinguished from dessert grapes; as outdoor or indoor; as black or white; or according to the time of ripening. But no matter what the system, there is nearly always some overlapping between different categories.

Vitis vinifera probably originated in Western Asia. It is one of the oldest of cultivated plants, having been known to the ancient Egyptians about 6,000 years ago. Wine may have been made from wild grapes in the Caucasus region in even earlier times. The Greeks and the Romans developed viticulture to a high degree and they introduced vines and wines into their colonies. Vineyards flourished throughout Romanized Europe. Even in favourable parts of Britain viticulture prospered until about the fourteenth century, when a decline set in which has been attributed to an unfavourable change in climatic conditions and the competition of imported wines from France. The dissolution of the monasteries during the sixteenth century no doubt hastened the downfall of British viticulture although some vineyards persisted until the eighteenth century. Recently there has been some renewal of interest in the possibility of successful viticulture in southern England and Wales.

Most of the dessert grapes eaten in the British Isles are imported from countries where the climate is favourable for their growth and ripening out of doors, for example, Spain and South Africa. However, the finest quality dessert grapes, for size, succulence, and flavour, are grown under glass in this country. Those most favoured by gourmets require artificial heat for fruit setting and ripening; examples are 'Muscat of Alexandria' which has amber-green fruits, and 'Gros Colmar' which has blue-black fruits. 'Black Hamburgh' (2) is easier to manage and is suitable for cultivation in a cold greenhouse.

'Chasselas Doré de Fontainebleau' (4), also known as 'Golden Chasselas', is suitable for growing in cold greenhouses and under cloches, and also in warm, sheltered situations out of doors, against a wall or a fence. There are many earlier ripening cultivars which can be grown in the open in mild localities and will ripen under cloches even in cold districts. 'Perle de Czaba' (3) was one of the first to be recommended for this purpose during the post-war period of increasing interest in outdoor grapes. Its medium-sized, yellowish fruits have a muscat flavour, but it is not a very reliable cropper. 'Pirovano 14' is a very early red or black grape of good flavour which can be relied upon to ripen out of doors even when the season is not favourable.

Dried Grapes. Dried grapes, known as raisins, sultanas and currants, are very important articles of commerce. Methods of production vary to some extent. In Spain, the finest dessert raisins are dried on the vines, by partially cutting the stalks of the bunches and removing the leaves to expose the grapes to the sun. Usually, however, the ripe bunches are cut and spread out to dry on floors exposed to the sunlight. Other important producing countries include Greece, South Africa, the United States, Australia and parts of Asia Minor. Although here placed with the Dessert Grapes, the varieties used for drying are usually wine grapes, because they are firm-fleshed and have a high sugar content.

Sultanas are rather small seedless raisins, produced in most of the countries where seeded raisins are also grown.

Currants are derived from a smaller-fruited black grape variety which has been grown in Greece for more than 2,000 years. Greece remains the chief commercial source, but some quantity is also produced in Australia.

TWO-THIRDS LIFE SIZE

1 GRAPE VINE, flowers, and leaves.

2 'BLACK HAMBURGH'   3 'PERLE DE CZABA'   4 'CHASSELAS DORÉ'

WINE GRAPES

Wine is made from grapes in many parts of the world for local use. France and Germany have long been the best known exporting countries for table wines — although the annual production of wine in Italy often exceeds that of France — but many countries, notably Spain, Portugal, and Yugoslavia, have increased their wine export trade in recent years. The fortified wines of Spain (sherry) and Portugal (port) are so well known that other producing countries have often called their fortified wines by the same names in order to compete in international markets.

Wine is produced by fermenting the juice crushed from grapes. Fermentation is a process by which yeast (a microscopic fungus) converts sugar into alcohol and carbon dioxide gas. In the making of most wines, the gas is allowed to escape, but in the special processes of making champagne and other sparkling wines, the gas is retained to form the characteristic bubbles. Yeasts occur naturally on the surface of the fruit, but artificially cultured wine yeasts are sometimes added to the juice. Fortified wines owe their high alcohol content to added brandy.

The flavour of wine is influenced by the characteristics of the grapes from which the juice is obtained, and also by the climate and the soil of the region and the expert knowledge of those who grow the grapes and make the wine.

About the middle of the 19th century, Phylloxera, an aphid which was probably introduced on American vines imported for trials, threatened to destroy the vineyards of Europe. Ironically, they were saved by grafting susceptible European varieties of *Vitis vinifera* (*see* p. 90) on resistant American rootstocks. The North American species, such as the Fox Grape (*Vitis labrusca*) and the Muscadine (*Vitis rotundifolia*), have a 'foxy' or musky flavour and are generally considered to be inferior to varieties of the European Grape Vine (*Vitis vinifera*). The latter is a vigorous climber, 50 to 60 feet high if left unpruned, but usually restricted by pruning. It climbs by means of forked tendrils, produced intermittently at two out of three vegetative nodes. Its leaves are 3 to 9 inches wide, long-stalked, palmately lobed and coarsely toothed. The small, greenish flowers have their petals joined at the tips.

There are hundreds of cultivated varieties of *Vitis vinifera* but relatively few of these are of outstanding importance.

1 **'Chardonnay'** or **'Pineau Blanc'** bears the rather small, roundish, yellowish-green fruits, from which white Burgundy is made. The same grape, grown in chalky soil in the more northerly district of Champagne, yields a very different wine, which derives its name from the region in which it is produced. Sparkling wines are made in other regions of France, and in other countries, but champagne reigns supreme.

2 **'Pineau Noir'** is a purplish-black grape, used traditionally for making red burgundy. Grown on the sunny, south-eastern slopes of the Côte d'Or, this grape yields red wines of distinction. In other regions, given different soil and climatic conditions, the same grape yields wines which rarely approach the flavour of burgundy, although they may be excellent in their own way.

3 **'Cabernet'**, a bluish-black grape, is one of a group of varieties used for making the famous red wines of Bordeaux that in Britain are usually called clarets.

4 **'Riesling'** is a collective term for a group of white (i.e. green) grape varieties, yielding the fine and sometimes very costly hocks of the Rhine wine districts. 'Hock' is a British term, believed to have been derived from the name of the port of Hockheim, through which many wines were exported to Victorian Britain. The rieslings have been widely planted elsewhere and their name has become better known in Britain through the availability of inexpensive riesling wines from Yugoslavia and Hungary.

TWO-THIRDS LIFE SIZE

1 'CHARDONNAY'   2 'PINEAU NOIR'
3 'CABERNET'   4 'RIESLING'

93

FIG, MULBERRY, AND POMEGRANATE

1 **Fig** (*Ficus carica*). Probably a native of western Asia, the fig has been grown for its fruit since very early times and is widespread in warm temperate and subtropical regions. It is sometimes self-sown and naturalized, even as far north as the British Isles. Fig seeds have been identified amongst Roman remains in Britain. Figs are eaten fresh, dried, tinned, or preserved. In addition to their food value (they contain about 50 per cent of sugar, when dried) they have mild laxative properties. ('Syrup of Figs' is a well-known medicinal preparation, included in the British Pharmaceutical Codex.) Fresh figs have a tender thin skin and are difficult to transport and keep in good condition. Consequently, they are best eaten in the regions where they are grown. With dried, tinned, or preserved figs, this difficulty does not arise and there is considerable world commerce in these commodities. The producing regions include Asia Minor, Greece, Italy, Algeria, Portugal, and California.

The fig is a deciduous shrub or small tree, up to about 30 feet high, belonging to the Mulberry family (Moraceae). Its leaves are palmately 3- or 5-lobed. Its mode of flowering is interesting — the flowers are borne inside the fruit, which is actually a multiple fruit consisting of the swollen fleshy axis. Pollination is done by the female fig wasp, which crawls through a small hole at the end of the fig to reach the flowers. The fig wasp does not occur in the British Isles, but the varieties of figs grown here are parthenocarpic, that is, they are able to fruit without being pollinated. The fruit is more or less pear-shaped and green, brown or purplish in colour, in different varieties. Up to three crops per year may be produced, in warm countries or in glasshouses; but out-of-doors in cool climates, even when trained against a warm wall, the fig seldom ripens more than one crop per year — the spring-formed fruits which mature in summer. Later-formed figs usually drop off, but even if they survive a mild winter they are apt to be attacked by mildew and become malformed. There are numerous varieties of figs, amongst the best-known being '*Brunswick*', '*Brown Turkey*', '*Black Ischia*', and '*White Ischia*'.

2 **Mulberry** (*Morus nigra*). The Common or Black Mulberry is probably a native of western Asia, but it was established in Europe in ancient times and was mentioned by Greek and Roman writers. It has never been of much commercial importance, because its fruits must be picked when fully ripe and are very juicy and easily squashed. When eating them, it is difficult to avoid staining hands, lips, and sometimes clothing with their bright purplish-red juice. Mulberries are delicious when eaten fully ripe and fresh from the tree. They can also be used for making mulberry wine and mulberry jam.

The botanical name of the genus *Morus* provides the root for the name of the family to which it belongs — the Moraceae. The Black Mulberry is a picturesque deciduous tree, up to 30 feet high, with a short, rugged trunk, branching crookedly to form a broad, irregularly dome-shaped crown. The dark green leaves, 3 to 9 inches long, are irregularly toothed and sometimes lobed. The unisexual flowers are joined together in a dense cluster (2A), their fleshy bases and perianth parts swelling to form a multiple fruit, green at first, then pink, and finally purple. The Mulberry is easily grown from cuttings, or by layering.

White Mulberry (*Morus alba*). This native of Asia is grown primarily for its light green leaves, which are the food of the common silkworm. Its white, pink, or purplish fruits are sweet, but rather insipid.

3 **Pomegranate** (*Punica granatum*). Said to be a native of western or southern Asia, the pomegranate is now widely cultivated in the tropics and sub-tropics and naturalized in the Mediterranean region, South America, and elsewhere. An ancient Latin name for the pomegranate was *Malum punicum* or 'apple of Carthage'. The fruit is eaten raw — although it is not to everyone's taste, having large numbers of seeds and a quantity of acid pulp in relation to the comparatively small amount of juicy flesh. In the East, the juice is used in cool drinks, which is perhaps the best way of avoiding the tedium of sucking it from around the seeds or swallowing the latter. The juice is also used for making a kind of wine and the seeds are used in conserves and syrups.

The pomegranate is a deciduous shrub or small tree, so distinctive in its botanical characteristics that it is often classified in a family of its own, the Punicaceae. It bears glossy, oblong-lanceolate, short-stalked, mostly opposite leaves. Its vivid orange-red flowers (3A) have 5 to 7 sepals, which persist at the apex of the fruit, 5 to 7 petals, and numerous stamens. The fruit is a hard, thick-skinned berry, crowned with the persistent calyx and the shrivelled stamens, and divided by walls of pith into several cells, each containing numerous seeds embedded in pink or crimson acid-sweet pulp.

The pomegranate is hardy enough to be grown out of doors in the British Isles, as a curiosity and for its ornamental flowers, but even when given the protection of a warm south or west wall, it rarely ripens its fruit here. There are several named varieties, with white, red and white, or red flowers, either single or double.

2A

2B

1

2

3A

3

TWO-THIRDS LIFE SIZE DETAIL × 2

1 FIG

2 MULBERRY 2A Female flower-head 2B Detail of flower

3 POMEGRANATE 3A Flowers

95

TROPICAL FRUITS OF THE AMERICAN CONTINENT

1 Pineapple (*Ananas comosus*). This is one of the most popular and delicious of tropical fruits. Native to South America, it is now widely distributed in the tropics and can also be grown in greenhouses in the temperate zone. The fruit is a multiple organ, formed by the coalescence of the fruits of a hundred or more individual flowers. Pineapples can be propagated by planting the crown of leaves which surmounts the fruit (1 and 1A), or small shoots which appear just below the fruit ('slips'), or others produced nearer the base of the stem ('suckers'). Of these, the last-named produce fruit more quickly, the time taken from planting to maturity varying from about 15 months in tropical lowlands to 20 months or more in the highlands. The leaves in most varieties of pineapple have sharp spiny edges (1) which hamper the movements of workers among the crop; varieties with smooth leaves have therefore been selected and are often planted. In commercial cultivation, the plants are sometimes given an application of plant hormone which induces fruiting; more uniform time of ripening is thus achieved which simplifies harvesting, and can to some extent be varied to meet seasonal demands. A crop can continue bearing for many years, but as the fruits gradually grow smaller, the usual practice is to replant after two or three years of bearing.

The popularity of pineapples as dessert fruit is mainly due to their high sugar content and attractive flavour; there is also a useful content of vitamins A and C. Export products include both fresh fruits and canned chunks or slices. The most important producing country is Hawaii, where pineapples are grown on a plantation scale and the cultivation methods are among the most scientific for any tropical crop. There is also production on a commercial scale in Malaysia, Australia, South Africa, and a number of other tropical countries, some of which also export.

2 Ceriman (*Monstera deliciosa*). Originating in the American tropics, this is a minor fruit and also a botanical curiosity, being the only genus of plants which has natural holes in the leaves. For this reason it is sometimes grown as an ornamental in greenhouses. The plant is an epiphytic climber which grows up trees and can form long aerial roots; it belongs to the same family as the Arum lily, and the part eaten is the spadix or cone-like fruit.

3-4 Anonaceous Fruits. There are a number of species of the family Anonaceae, which is native to the American tropics. Their fruits are still relished more in that region than anywhere else though some, like the cherimoya, sweet sop, and bullock's heart, have also been fairly widely planted in tropical Asia. Collectively, the fruits of this family are sometimes known as 'custard apples' from the custard-like flavour of many. This name is, however, sometimes also applied to one or other of the individual species, and there is similar confusion in the use of the common English names in different countries. All the species are small trees, and superior varieties are often propagated by budding. The fruits rarely appear in the markets of the temperate zone. Botanically, the flowers bear numerous separate carpels on an elongated receptacle and these become aggregated into a multiple fruit or 'syncarp'.

Sour Sop (*Anona muricata*) (3). Its very characteristic fruit, with rows of soft spines on the green rind, may weigh up to 8 lb. The flesh is white and fibrous, more acid and less sweet than most in the group.

Cherimoya (*Anona cherimolia*) (4). The fruits of this very popular member of the group have a pineapple flavour. They have a scaly surface such as is common to most of the genus. This species is adapted to the tropical highlands, unlike the others, which are lowland plants.

Sweet Sop (*Anona squamosa*). According to some people, this is the true 'custard apple'; it is also known as the sugar apple. The sweet, custard-like flesh has a pleasant aroma, and the fruit is particularly popular in the West Indies.

Bullock's Heart (*Anona reticulata*). Its common name is due to its sometimes heart-shaped fruit of buff or reddish-brown colour. The flesh of the fruit is more solid than some, granular and sweet.

Ilama (*Anona diversifolia*). It resembles the cherimoya but can be grown in the lowlands. It is particularly appreciated in Mexico and Central America.

Soncoya (*Anona purpurea*). This tree, grown almost exclusively in Mexico and Central America, has particularly large fruits.

PLANTS AND BRANCHES × ⅛   *FRUITS* × ½   *FLOWER DETAILS* × 1
1 PINEAPPLE flowering and fruiting plants   1A Fruit   1B Flower detail
2 CERIMAN Fruiting branch
3 SOUR SOP Fruiting branch   4 CHERIMOYA Fruiting branch   4A Fruit   4B Flower detail   97

TROPICAL FRUITS OF THE AMERICAN CONTINENT

1 Passion Fruit or **Purple Granadilla** (*Passiflora edulis*).
A perennial climbing plant, originally native to Brazil
but now widely planted in the tropics, it is also
sufficiently hardy to be grown in some Mediterranean
countries. Propagation is by seed or cuttings; in
commercial cultivation, the plants are generally provided with wires stretched between stakes or with a
trellis to support them. The fruit, about the size of an
egg, is purple when ripe, but on ripening quickly
assumes a wrinkled appearance which makes it look
old when in fact it is at its best. The sweet juicy pulp is
inseparable from the small blackish seeds which have
to be swallowed with it if the fruit is used fresh.
Bottled passion fruit juice, which is free of seeds, is a
popular drink mainly in the countries where it is
produced.

2 Giant Granadilla (*Passiflora quadrangularis*). Less
widely grown than the passion fruit, it has very showy
and attractive flowers (2A) and the largest fruits of all
the granadilla group. It is more confined to the tropics
that the passion fruit, and is again a climber, with a
useful life of 5 – 6 years. The fruit is green or greenish-
yellow and its taste more insipid than the passion
fruit; in the unripe state it is sometimes boiled as a
vegetable.
There are also several other members of the granadilla
group which are occasionally grown for their fruits
but have not achieved widespread popularity. These
include the 'water-lemon' or 'yellow granadilla'
(*Passiflora laurifolia*) and the 'curuba' (*Passiflora
maliformis*), neither of which has spread much outside
the American tropics.

3 Sapodilla (*Achras sapota*). It is still grown most widely
in its native habitat of Central America, but is also
planted elsewhere in the tropics. The tree is of medium
height (15 – 60 ft.) and of handsome shape. It is most
commonly grown from seed, but some named varieties
are propagated by grafting. The fruits are brown and
filled with a luscious pulp in the middle of which are
embedded the black shining seeds, which are not
eaten. The fruit is only palatable when fully ripe and
then has a particularly attractive flavour rather
resembling brown sugar; it is considered one of the
best fruits of the American tropics.
The sapodilla tree also yields another edible product,
chicle gum. This is obtained by making incisions in
the bark, from which flows the milky latex which is
collected and then coagulated by heating. Chicle gum
provides, with the addition of various flavouring
substances, the basis for chewing-gum manufacture;
it is exported in substantial quantities from Mexico,
British Honduras, and other Central American
countries.

4 Guava (*Psidium guajava*). This has spread from the
American tropics to become one of the most commonly
planted tropical fruits. It is a small tree growing to
about 30 ft. high and can readily be recognised by the
characteristic bark of the younger branches which is
smooth and reddish-brown, and peels off in thin
flakes. Propagation is by seed or grafting. The fruits
are green, turning light yellow when ripe. The best
varieties are juicy, but the pulp is full of small seeds
and the taste is somewhat sharp if eaten raw. The
fruits are often stewed or made into tarts, while
another typical product is guava jam or jelly which
seems to have originated in the West Indies and is
sometimes exported from there. Some canned guava
also enters world trade. Guavas are notable for their
high content of vitamin C, which in many varieties is
several times greater than in the citrus fruits.
A number of other species of *Psidium* are occasionally
cultivated. Perhaps the best known of these is *Psidium
cattleianum*, the 'strawberry guava', which has fruits
smaller than the ordinary guava and of a reddish-
purple colour with a good flavour; it is somewhat
more hardy than the ordinary guava.

TWO-THIRDS LIFE SIZE (VINES AND Fig. 2 × ¼)

1 PASSION FRUIT   1A Flowering vine   2 GIANT GRANADILLA   2A Flowering vine
3 SAPODILLA   3A Flowers   4 GUAVA   4A Flower   4B Immature fruit   99

TROPICAL FRUITS OF INDIA AND MALAYSIA

1 Mango (*Mangifera indica*). One of the most popular and characteristic fruits of the tropics, the mango originated in the Indo-Burmese region. It is now most widely grown in India where it is the best loved fruit, but it is also planted in almost every tropical country and sometimes slightly beyond the limits of the tropics, as in northern India, Egypt, Natal, and Florida. It flourishes best below 2,000 feet altitude and is tolerant of a wide range of soils and rainfall, though not suited by a continuously wet climate as heavy rainfall reduces pollination.

The mango is a medium-sized to large tree of rather variable height. The narrow dark green leaves form a characteristically dense mass of foliage, casting a heavy shade which makes the tree useful as a firebreak but incidentally provides a favourable habitat for mosquitoes. Fruiting begins by about 6 years from planting, but yield declines after the trees are about 40 years old. There is a tendency to biennial bearing, with exceptionally heavy crops in some years. Flowering takes place usually between January and March in the northern hemisphere, and in June to August in the southern. The inflorescence (1B) produces clusters of small numbers of fruits, each hanging down on a stalk which in some varieties can be very long. The yellowish-green skin of the fruit is not eaten. Inside it is the orange-coloured edible flesh surrounding the central seed (stone). It is difficult to eat the fruit without getting sticky.

Many mangoes are planted from seed, but such trees cannot be relied on to come true to type, and inferior mango fruits are not attractive because the pulp can be stringy and the taste reminiscent of turpentine. Vegetative propagation of good trees is therefore much used, usually by the process of approach grafting or inarching, though budding is also practised. The produce of such trees is undoubtedly one of the world's best fruits. There are many famous and popular varieties, among the best known being 'Alphonso' and 'Mulgoa' in India, and 'Julie' and 'Peter' ('Bombay') in the West Indies. Mango fruits contain 10 – 20 per cent of sugar, are an important source of vitamin A, and contain some of vitamins B and C. They are mainly eaten fresh, but are also made into preserves and sometimes canned; and unripe fruits are used in making mango chutney. A single fruit weighs from 6 oz. to 1½ lb.

By far the most important producing country is India, where there are estimated to be 2 million acres of mangoes, producing about 5 million tons of fruit a year. Small numbers of mangoes are flown from the tropics to the markets of the temperate zone at a high price, but no considerable trade has yet developed.

2 Rambutan (*Nephelium lappaceum*). It is a fruit native to Malaysia and very popular in the south-east Asian countries. It belongs to the same family as the litchi (*see* p. 105), the Sapindaceae. The tree is generally grown from seed or sometimes by bud-grafting which ensures desirable characters, and is large and handsome, reaching a height of up to 60 feet. The clusters of fruits (2A) are of most striking appearance; each fruit is about the size of a plum and is covered with soft spines which are bright red or sometimes yellow. The part eaten is the white fleshy aril surrounding the single seed. This has an attractive and refreshing sweet-acid taste and is usually eaten raw, but may also be stewed. The rambutan is limited to tropical lowlands and requires a high rainfall. In the south-eastern Asian countries it is very commonly planted in gardens and small orchards, and has very much the same distribution as the durian. There are a number of named varieties in this region; it is little grown in other parts of the world.

A related species *Nephelium mutabile*, the pulasan, is planted on a smaller scale in the same region, especially in western Java. The fruit in this case is covered by short blunt red or yellow tubercles. Its popularity is more local than that of the rambutan.

3 Mangosteen (*Garcinia mangostana*). This is another native of Malaysia and requires a hot humid climate. It is a slow-growing tree which may reach 45 feet in height. The fruit is a berry of brownish-purple colour, surmounted by the calyx which remains attached. Inside the tough rind, the pulp is divided into segments which can be eaten individually; the flesh is whitish and soft and each segment may contain a few small seeds. The fruit is best eaten quite fresh, and has a delicious somewhat treacly taste. The mangosteen is generally admitted to be one of the most attractive of tropical fruits; but, although individual trees may be seen in many tropical countries, it is nowhere planted on a large scale. This is perhaps because of difficulties of propagation (the seed does not germinate very well, and vegetative propagation is not easy), and because of its slow growth which means that bearing may not even begin until it is about 15 years old. A mature tree may bear about 500 fruits a year.

FRUITS × ⅔   BRANCHES × ⅛   FLOWER DETAILS × 3

1 MANGO fruit   1A Fruiting and flowering branches   1B Flower detail
2 RAMBUTAN fruit   2A Fruiting branch   2B Flower detail
3 MANGOSTEEN fruit   3A Fruiting branch   3B Young fruit

SOME REGIONAL TROPICAL FRUITS

1 **Carambola** (*Averrhoa carambola*) is a native of Indonesia; small numbers of trees have been planted in other parts of the tropics. The plant is a small tree from 20 to 35 ft. high, usually propagated by seed but sometimes by grafting in order to perpetuate good strains. The fruits, which are 3 to 5 inches long, ribbed, and of a very attractive yellow colour, are borne in great profusion on the plant. They are very juicy but variable in taste; some trees bear sweeter fruits, and some rather acid ones. The fruits are used in fruit salads, jellies, tarts, and preserves, and in making a drink. They do not enter into world trade but are consumed locally, principally in south-east Asia. A related species, *Averrhoa bilimbi*, is sometimes cultivated in the same region and the fruits are used in a similar way and in curries.

2 **Durian** (*Durio zibethinus*). This is one of the most notorious of tropical fruits, for a peculiar reason. Its taste and smell have a sewage-like quality which many people find so offensive that they would be disgusted to consume the fruit; on the other hand, durians are much esteemed by the people of south-east Asia, and some foreigners who have acquired the taste for them consider them to be one of the most delicious tropical fruits. The tree, which is native to Malaya, is handsome and large, up to 100 ft. tall. It is propagated by seed, and fruiting begins at about 7 years old. The fruits are large, usually weighing from 6 to 8 lb., covered with spines, and a dull yellow in colour when ripe. The fruits, when opened, expose a creamy pulp, which is the part eaten and which botanically consists of the arils surrounding the seeds. The pulp is eaten raw and the fruits must be fresh, as their quality deteriorates quickly. Fruit is provided not only by wild trees and garden trees, but also by whole orchards of durians which are planted widely in south-east Asia; the total acreage under these trees in Malaysia, Indonesia, and Thailand must be considerable, though it has never been measured. In these and neighbouring countries the fruit is sold in great numbers in local markets. But there is no export demand and in the rest of the tropics the fruit has never attained enough popularity to be planted except as a curiosity.

3 **Akee** (*Blighia sapida*). This fruit of West African origin has also long been grown in the West Indies. It was introduced to Jamaica in the late eighteenth century, and is now particularly associated with that island where it has its greatest popularity. The tree is medium-sized and begins to bear from about 5 years old. The fruits are about 3 inches long and red when ripe. They open naturally, exposing usually three seeds in each fruit. Each seed is surrounded by a fleshy cream-coloured aril which is the part eaten. The aril can be eaten raw but is usually fried or boiled. The pink tissue joining the aril to the seed must not be eaten as it is highly poisonous. Since under-ripe and over-ripe fruits can also be dangerous, only ripe fruits which have opened naturally but are still in a fresh condition should be used.

1A

1

2

3A

3

TWO-THIRDS LIFE SIZE

1 CARAMBOLA   1A Flowers
2 DURIAN
3 AKEE   3A Flowers

103

CHINESE AND JAPANESE FRUITS

1 Litchi (*Litchi chinensis*). This is also sometimes spelt in English as 'litchee' and 'lychee'. It is a member of the family Sapindaceae, and is botanically related to the rambutan (*see* p. 100). The fruit has long been appreciated in its native China, where it is mainly planted in the tropical and subtropical south. It has also been planted on a lesser scale in many tropical countries and a few subtropical areas such as Florida, but does not fruit well in very humid tropical areas. This medium-sized tree (30 to 40 feet) bears bunches of fruits, the size of plums, with a rough warty rind of an attractive pinkish crimson colour. The edible part is the pulp which is actually the aril of the single glossy-brown seed; it is a translucent white jelly-like substance with a sweet acid flavour which makes it very acceptable as the last course of the traditional Chinese dinner. Litchis are most usually eaten fresh, but are also sometimes canned for export or preserved in syrup; litchi 'nuts' can be prepared by drying the fruit, the pulp then taking on a nutty raisin-like flavour. Most of the trees in cultivation are seedlings, but superior varieties can also be propagated by marcottage (aerial layering).

2 Persimmons or date plums are fruits of the warm temperate zone, of which three species are utilised. The Japanese persimmon (*Diospyros kaki*), which is the one shown in Figure 2, is by far the most widely cultivated. It is an important fruit in its native region of China and Japan, is popular in the southern United States to which it was introduced about 1870, and is also cultivated in the south of France. The trees are deciduous and are propagated by seed or grafting. The fruits are 2 – 3 inches or more in diameter and yellow to red in colour, thus having much the size and appearance of a tomato. They are eaten fresh, cooked, and sometimes candied. In the south-eastern United States the trees are mostly planted in home gardens, but in California there is a small commercial acreage. The American persimmon (*Diospyros virginiana*) is native to the southern United States and will fruit as far north as the great lakes. The fruits are smaller than the Japanese persimmon and dark red to maroon in colour. The species is not often planted, but considerable quantities of the fruit are picked from wild trees. This species is often used in America as a rootstock on which to graft the Japanese persimmon.

A third species, *Diospyros lotus*, is grown to a lesser extent in the Far East and Italy.

3 Loquat (*Eriobotrya japonica*) is also sometimes known as Japanese medlar; it belongs to the Rosaceae, the same botanical family as the apples, and is one of its few cultivated subtropical representatives. The most important acreages are in its native region of China and Japan and in north India; it has also been quite widely planted in Mediterranean countries. It can be successfully grown in the tropics at medium elevations (above about 3,000 ft.). The tree is a small evergreen, usually propagated by seed or grafting. The leaves are narrow, and dark green on the upper surface with a characteristic lighter woolly under-surface (3B). It produces clusters of white flowers in autumn, and fruit in the spring. The yellow pear-shaped fruits are the size of crab apples and have a sweetish acid flavour, but the quality is best in selected varieties. The fruits can be eaten fresh or stewed and are also made into jam or jelly. In Bermuda they are used in making a local type of liqueur.

FRUITS AND FLOWERS LIFE SIZE   FRUITING BRANCHES × ¼

1 LITCHI fruits	1A Flowers	1B Fruiting branch
2 PERSIMMON fruit	2A Flowers	2B Fruiting branch
3 LOQUAT fruits	3A Flowers	3B Fruiting branch

DATE AND PALMYRA PALMS

1 Date Palm (*Phoenix dactylifera*). This is an important crop plant of very ancient origin, with records of cultivation in the Middle East dating back to at least 3000 B.C. It is a tall palm, growing to about 80 feet high, and the trunk is well protected by being completely covered in the persistent leaf-bases of long-dead leaves. The male and female trees are separate, and it is only necessary for the grower to plant one male tree to 50 – 100 females, or about one male per acre. Fruit set can, however, be improved by artificial pollination, which most growers achieve by cutting off clusters of male flowers (1D) and fixing them among the branches of the female flower-bunch (1B). Though dates can be grown from seed, purity of variety in commercial cultivation is ensured by propagating them from offsets or suckers which are produced by the tree during the earlier part of its life; surplus suckers are normally pruned away to leave the palm with a single stem. Palms may begin to bear about 4 – 5 years after planting; they are in full bearing by about 15 years, and may continue fruiting up to about 80 years of age. A good fruit bunch (1A) will have about 40 strands of fruit with from 25 to 35 dates per strand. An average yield is about 100 lb. of fruit per tree per year, with good trees yielding 150 lb.

The date palm is essentially a crop of dry sub-tropical areas and is usually grown with irrigation. Though its cultivation does penetrate into a few arid parts of the tropics, it is not grown in humid climates where the yield is less and the fruit of poor quality. In the Old World, dates are grown from Morocco in the west to India in the east. Iraq is much the most important producing country, and is followed by Saudi Arabia, Algeria, Iran, and Egypt. Date palms have been taken to the New World and are successfully grown on a small scale in the drier southern parts of the United States, such as California and Arizona.

Dates are often grown in desert oases, and are an important part of the diet of many Arab populations. The kinds of fruit are generally classified into three types. The first is the soft dates, which are grown on a large scale in Iraq and exported, mainly to other parts of the Arab world. Soft dates are often sold in pressed masses, and are eaten raw or used in confectionery. The second class is the semi-dry dates; in this class may be placed the variety 'Deglet Noor', the world's most popular date. These dates (1G) are often sold boxed, with the fruits still attached to their strand. Imports of these dates into Europe come largely from the North African countries. The third class comprises the dry dates, which can be kept for a long time and so tide over the periods when the palms are not bearing. They are quite hard and can even be ground into flour, but can also be softened by steeping in water before eating. They are very important articles of diet in Arab countries, between which there is a considerable trade, but are not much exported to other parts of the world. The chief nutritional value of dates is in their high sugar content, which varies from about 60 per cent in the soft dates to as much as 70 per cent in some dry types. They also have some content of vitamins A, B_1 and B_2, and nicotinic acid.

In addition to the fruit, date palms, like many other palms, provide a food supply in the sugary sap which can be obtained by tapping the crown of the plant. This can be fermented to make palm wine or 'toddy', or it can be boiled down to provide sugar. Tapping must be done with moderation if the yield of fruit is not to be seriously reduced; but it is a way of making use of senile palms which are past the age for good fruit-bearing.

2 Palmyra or **Borassus Palm** (*Borassus flabellifer*). This is a palm of the drier tropical regions which grows wild in parts of south India, Ceylon, Burma, and tropical Africa, but in Asia is also often planted. In these dry areas it takes, in a sense, the place of the coconut palm which needs a higher rainfall. The palmyra is a tall palm of much the same height as the date palm, with fan-shaped leaves, and again has separate male and female trees. A female tree may bear about 200 nuts a year which contain a sap used as a refreshing drink; the soft kernel of the young fruit is also eaten, and germinated nuts have an enlarged fleshy embryo which is used as a vegetable. The palm is perhaps even more widely planted in order to tap the spadix of the inflorescence for the sweet sap, which is used for making toddy or sugar. Its use for sugar production is described on page 16 and it is illustrated in more detail on page 17. As with the date palm, the trunks of old trees are much used for local building purposes.

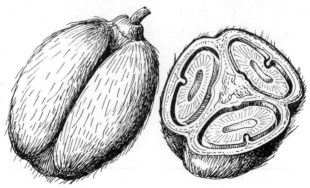

FRUIT OF PALMYRA PALM ($\times \frac{1}{3}$) Approximately

FRUITING SPADICES × 1/20   FLOWERING SPADICES × ⅛   FLOWER AND FRUIT DETAILS × 1

1 DATE PALM (small scale)

1A Fruiting spadices   1B Female spadix in flower   1C Detail of female flowers
1D Male spadix in flower   1E Detail of male flowers   1F Ripe fruit   1G Dried fruit

2 PALMYRA PALM (small scale)

107

BANANA

The bananas belong to the genus *Musa*, but as the cultivated kinds are sterile hybrid forms, they cannot be given exact species names. These forms with edible seedless fruits were derived by man in prehistoric times from wild bananas which grow in the region stretching from India to New Guinea; they are now cultivated in all parts of the tropics. The plant, though often referred to as a 'tree', is really a giant herb, whose stem is composed of the overlapping bases of the leaves above; it is typically from 10 to 30 feet high. The leaves are very large and handsome (1 and 1c) but because of their large size, plants are liable to be blown over and the leaves become torn by the wind into strips which finally give a tattered appearance. Propagation is by planting out suckers which arise from the rhizome below ground, or sometimes by planting pieces of the rhizome itself which contain buds.

Within a year after the sucker has been planted, the flowering stem will emerge at the apex of the plant and will gradually bend over to hang downwards (1B). At its end this stem carries the sterile male flowers protected by large red bracts which are a conspicuous feature until they wither and fall away. Higher up the stem are borne the groups of female flowers, from which the seedless fruit develop without fertilisation. The whole fruit bunch consists of a number (usually about 9 – 12) of half-spirals of fruits which are clearly separated from each other and are called 'hands'. Each hand typically contains about 12 – 16 individual fruits, which are sometimes called 'fingers'. Thus the whole bunch may consist of about 120 – 200 individual fruits or bananas and weigh 50 – 80 lb. The skin of the fruit is green when unripe and turns yellow when ripe. The skin is easily peeled off to reveal the edible flesh (1D), which in most varieties is white with a tinge of yellow.

After fruiting, the stem which has borne the inflorescence dies and is usually cut down. Meanwhile however the rhizome has thrown up other suckers which in turn flower and die, and in this way the plant goes on fruiting indefinitely, the whole group of stems arising from one rhizome being known as a 'stool' or 'mat' of bananas. In commercial cultivation, fields of bananas are most usually kept in existence for from five to twenty years before replanting is undertaken, but many small farmers in the tropics maintain patches of bananas for fifty or sixty years, and the plants can live much longer than that.

Banana fruits are used in several different ways. The most widely known kinds, and the only bananas imported into temperate countries from the tropical producing areas, are the dessert bananas which because of their high sugar content (about 17 – 19 per cent) have a sweet taste and are eaten raw. The export trade is dominated by a few varieties which have proved able to stand the period of transport. Most consumers do not discriminate, or even know the difference between such varieties as 'Gros Michel', 'Lacatan', and 'Robusta'. 'Canary' bananas are borne on a dwarf banana plant and can be grown in some subtropical climates, as in the Canary Islands and Israel; they are distinguished by the fruit being somewhat smaller and thinner-skinned than those already mentioned.

The banana export trade is a highly organised one, using specially equipped ships whose holds can be kept at a constant temperature. The bananas for this trade are cut in the green stage before they are fully ripe, and are finally ripened in store in the importing country. The largest single exporting country is Ecuador, whose chief customer is the United States. There are large company-owned estates in several Central American countries, and much production by small farmers in the West Indies; exports from the Caribbean area go both to the United States and Europe. Europe is also largely supplied from West Africa, with the Ivory Coast and Cameroon as leading exporters. South Africa and Australia are able to grow bananas in the warmer parts of their own countries; New Zealand imports from Fiji and other Pacific Islands. Besides this export trade, which includes many other producing countries of lesser importance, there is in most parts of the tropics a large local consumption of sweet bananas including very many delicious varieties which are not suited to the export trade. Examples are the very short fruits commonly known as 'Lady's Fingers' of which one kind is illustrated (2), and the red-skinned type, which may also have an orange tint in the flesh (3). Besides the trade in fresh fruit, little commercial use is made of bananas. There is however a small production of dried bananas, and of banana flour which is more digestible than the cereal starches.

The second main use of bananas is for cooking; this is almost wholly confined to local consumption within the tropics, mainly by the growers themselves. Bananas used for cooking have a higher starch and lower sugar content than dessert bananas, and are picked when their flesh is too hard to be eaten raw. They may be of varieties which would sweeten if left to ripen, or of others which never reach a high sugar content. Cooking bananas are sometimes called 'plantains', but this word can lead to confusion as it is used in some countries to describe certain types of dessert bananas. Many different varieties are grown, some specially for steaming and some for roasting. Cooking bananas are the chief staple food of some millions of people in East Africa, especially in southern Uganda and parts of Tanzania, but are also a minor crop in many other tropical countries.

A third use of bananas is for making beer, especially in East Africa where special varieties are grown for the purpose. A crop closely related to the banana, called 'ensete' (*Ensete ventricosum*), is a major food crop in southern Ethiopia, where both the rhizome and the inner tissues of the stem are cooked for food; the fruits are small and contain seeds.

PLANTS SMALL SCALE   DETAILS × ¼
1 BANANA plant   1A Young plant
1B Inflorescence   1C Leaf detail   1D Ripe fruits
2 'LADY'S FINGERS'   3 RED BANANAS

COFFEE, CHICORY, AND DANDELION

Coffee, the single most valuable agricultural export of the tropics, is of very great economic importance in world trade. The beans, roasted, ground, and brewed in hot water, provide the stimulating non-alcoholic drink used by one-third of the world's population. Commercial supplies of coffee come from more than one species of the plant.

1 **Coffea arabica** (Arabica coffee) supplies the largest quantity and the best quality of coffee beans. The small tree is pruned, on one of a number of systems, to keep the height within reach of the pickers and to encourage regular bearing. The economic life of the trees may not be more than about 30 years. The flowers (1A) are white and sweet-smelling, producing green berries which turn red when ripe. Each berry contains two beans, or occasionally only one, which are the commercial product of the plant. To produce the best quality beans — known in the trade as 'mild' coffee — the berries are opened by a pulping machine and the beans fermented for a short time in water and then sun-dried; this leaves the beans still in the inner or 'parchment' skin (1B). Much of the crop is exported in this form. A simpler process of sun-drying the whole berries and then removing the dried pulp from the beans in a hulling machine is practised in some countries; this produces 'naked' beans of lower quality, known as 'hard' coffee. Arabica coffee is grown mostly in the American tropics, where Brazil is by far the largest producer, her exports being generally of the 'hard' type. 'Mild' coffees are exported from many American countries, of which Colombia is the most important; the 'Blue Mountain' coffee of Jamaica is one of the world's best for quality, and very good coffees are also exported from some Central American countries such as Costa Rica and El Salvador. In the Old World, arabica coffee is attacked in the lowlands by a serious leaf disease and consequently cultivation has been concentrated in highland areas. Some Old World highland coffees, such as those of Kenya and of Mysore in India, have a very high reputation for quality.

2 **Coffea canephora** produces the other main type of coffee, known as 'robusta'. The tree is longer-lived and has larger leaves than arabica coffee; the beans (2) are smaller, generally prepared by dry-hulling, and of lower quality and price. The yield however is greater than from arabica, and as the trees are resistant to leaf disease they can be planted in the lowlands of the Old World. Robusta coffee is mainly produced in East and West Africa.

A third species, *Coffea liberica*, is very robust but produces beans of still lower quality. It is grown for local consumption in a few countries such as Malaysia and Guyana. Other species of coffee are now rarely grown.

3 **Chicory** (*Cichorium intybus*) is a well known substitute for coffee, often used blended with the latter, especially in liquid 'coffee extract'. It gives a bitterness to the beverage, which some people find refreshing. The part used is the root (3B), which is chopped, roasted, and ground. Varieties grown for the purpose have larger roots than some of the salad chicories (*see* p. 151), but their blanched leaves and young shoots can also be used for salads or as a cooked vegetable. '*Witloof*', which has shallowly lobed or toothed leaves, is used for both purposes. '*Brunswick*' has deeply cut leaves, resembling those of many dandelions.
Chicory, a member of the Daisy family (Compositae), is a perennial herb, about 3 feet high, with bright blue flower-heads (3A).

4 **Dandelion** (*Taraxacum officinale agg.*). The roasted and ground roots are used for making 'dandelion coffee', which is said to be almost indistinguishable from real coffee, and possesses tonic and stimulant properties yet lacks the possibly injurious substance, caffeine. The ground root (4B) is sometimes blended with true coffee or cocoa. The leaves can be used in salads, preferably blanched, like chicory and endive. The flower-heads (4A) are used for making dandelion wine. Wild plants can be used for all these purposes, but there are improved varieties in cultivation.
Taraxacum officinale is a perennial herb belonging to the Daisy family (Compositae). It is an aggregate species: that is, botanists have divided it into many species. Plants growing in different areas will show considerable variation in the shape of the leaves and the way in which they are toothed or lobed, and also in the form of the bracts at the base of the yellow flower-head.

1A

1

1

1B

2

4A

3A

4

4B

3

3B

TWO-THIRDS LIFE SIZE   *PLANTS* $\times \frac{1}{8}$

1 COFFEE fruits   1A Flowers
1B 'ARABICA' COFFEE berry (in section) and beans   2 'ROBUSTA' COFFEE beans
3 CHICORY plant   3A Flower head   3B Root   4 DANDELION plant   4A Flower head   4B Root

TEA AND COCOA

Tea was introduced from China into Japan about 1000 A.D., and became an important beverage in Europe in the 17th century. Nearly half of mankind now drinks tea. Cocoa, introduced from America to Europe in 1526, is a unique beverage because of its high food value.

1 Cocoa (*Theobroma cacao*). The plant is a small tree originating in the American tropics and strictly limited to the inner tropics, but now widely dispersed in this belt, the chief production being in West Africa. Fairly high rainfall and a good soil are necessities for this crop. Propagation is most commonly by sowing seed in a nursery, from which the seedlings are planted out about 6 – 8 months later; but in some countries outstanding trees are propagated by taking cuttings which are rooted in a special propagating bin. The trees begin to bear some crop at about 4 years old and may live to 80 years or more. Cocoa plantations are often interplanted with various taller trees to give shade to the crop, but recent research has shown that the highest yields can be obtained by a generous use of fertilizers without shade. The tree has the unusual habit of bearing its flowers, and subsequently its pods, on the main trunk (1 and 1B) as well as on the branches. Cocoa pods when mature are yellow in some varieties and red in others. When opened they disclose a mass of beans surrounded by white mucilage (1A). Beans and mucilage are scooped out and fermented; small growers do this by piling them in a heap covered with banana leaves, and on large estates it is done in wooden 'sweat-boxes'. The mature fermented beans are dull red in colour (1C) and must next be dried in the sun or in artificial dryers before they are ready for export. The main exporting region is West Africa, where the chief producing countries are Ghana, Nigeria, and the Ivory Coast; smaller exports, some of them of special quality, are made by a number of American, Asian and Pacific countries.

Cocoa manufacture takes place almost entirely in the importing countries. The processes include shelling, roasting and grinding the beans. Cocoa beans contain 50 – 57 per cent of a fat called cocoa butter. In the manufacture of cocoa powder for drinking, this fat is largely removed; but in making chocolate, extra cocoa butter is added as well as other substances such as sugar and milk. Chocolate has not only an attractive taste, but also a high fat content which makes it a very concentrated energy source. Cocoa is a mildly stimulating drink because of its content of caffeine and a related alkaloid theobromine.

2 Tea (*Camellia sinensis*) is a product manufactured from the leaves of a small tree which grows wild from India to China. The 'China' types are smaller trees with narrower leaves than the 'Indian' types. Tea can be grown from the equator up to latitudes as high as the Black Sea coasts of Russia, some provinces of northern China, and Japan. The crop revels in high rainfall, and is almost unique in giving particularly good yields on very acid soils. The plants are most commonly propagated by seed sown in a nursery, but cuttings can also be rooted. Trees for plucking are regularly pruned to a low bush shape which encourages maximum leaf production; but the few trees needed for seed-bearing are allowed to grow to their full height. Plucking is usually done by hand (2) at intervals of one week upwards; in some climates the growth is very seasonal. The picker usually takes the bud and two terminal leaves from the end of each shoot (2B), but for some teas the bud and three leaves are taken, giving a higher yield but a poorer quality product. Machines have been invented for removing the leaf from the bush, but are as yet in use only when labour is scarce. Some of the best teas come from high altitude areas such as Darjeeling in India, while tea from the plains is often of 'common' quality. The plants reach a full yield at about six years and may live to over 50 years.

The leaf after plucking has to be rapidly processed in a factory, such as has been built on most tea estates. The usual processes include withering, rolling, fermentation, drying, sifting, and grading. During these processes the leaf is broken up and turns from green to black; but in some countries, especially of the Far East, there is a demand for green tea which is produced by heating the leaf at an early stage to prevent fermentation. Black tea is produced in a number of grades, of which 'orange pekoe' is perhaps the best known; 'broken orange pekoe' is of higher quality, and 'souchong' of lower. Tea is traditionally exported in the familiar tea-chests lined with tin foil. China is the greatest tea-producing nation, but consumes most of what it grows. The export of 'China' teas has long been surpassed by the 'Indian' type teas, which are mainly produced in India and Ceylon, but now also elsewhere, notably in the East African countries. Of the importing countries, tea is most popular in Russia, Britain, and Australia. Tea contains tannin which gives the beverage its 'body', and caffeine which makes it stimulating.

Maté, or in full 'yerba de maté' (*Ilex paraguensis*) is another important tropical beverage crop. The leaves of this small tree are picked, dried, and ground by simple processes, and used with hot water to make a tea-like drink. The beverage contains a small amount of caffeine and is mildly stimulating. Maté is produced chiefly in Brazil and Paraguay, but the drink is popular in many other countries of South America. Little is exported to other parts of the world.

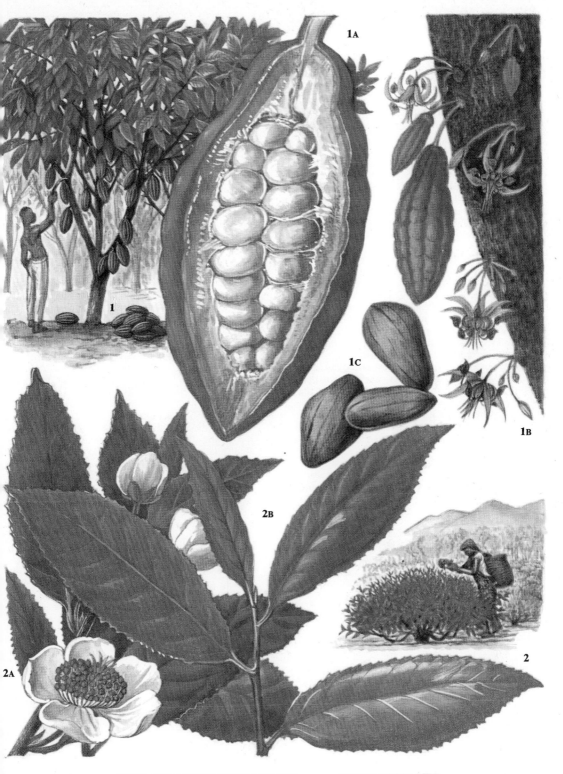

TREE AND SHRUB SMALL SCALE   *DETAILS LIFE SIZE*

1 COCOA TREE

1A Unripe fruit   1B Flowers and immature fruits   1C Fermented Cocoa beans

2 TEA SHRUB

2A Flowering shoot   2B Tea leaves

TROPICAL VEGETABLE FRUITS

1 **Papaya** (*Carica papaya*) is also commonly known as 'pawpaw' (sometimes spelt papaw). A native of tropical America, it is now widely planted all over the tropics. The plant, which is grown from seed, is a herbaceous 'tree', normally unbranched (1A) and extremely fast-growing with large handsome leaves. The male flowers (1C) and the female flowers (1B) are usually borne on separate trees, though some hermaphrodite trees occur. As soon as the sex of the plants can be determined, only one male tree is left as a pollinator to anything from 15 to 50 females. Bearing begins in the first year but falls off after 3 – 4 years, when the plants are usually replaced.

The fruits are much the size and shape of a somewhat elongated melon; some varieties remain green when ripe, but in most the skin turns yellow or orange. The succulent flesh is pinkish or orange, the mass of seeds being enclosed in the central cavity of the fruit (1). Ripe fruits contain about 7 – 9 per cent of sugars, and are particularly valued as a breakfast fruit, usually eaten in slices with added sugar and lime juice, and in fruit salads. In some equatorial countries they are in season all the year round, but in other tropical climates the fruiting season is more limited. The unripe fruits are sometimes boiled as a vegetable and may be served with white sauce like a marrow.

The papaya plant also produces an enzyme, papain, which has the property of breaking down protein. This is produced on a commercial scale by lancing the surface of the fruits, collecting the white latex which exudes, and drying it down to a powder. Papain is used in brewing, and in the preparation of canned meats, and sometimes medicinally; it is also used in various industrial processes unconnected with the food trade. On the domestic scale, papaya leaves are sometimes wrapped round pieces of tough meat with the object of tenderising them before consumption.

Mountain Pawpaw (*Carica condamarcensis*) has similar but smaller fruits which need to be cooked before eating or may be made into jam; its virtue is that it can be grown at higher altitudes in the tropics than the papaya.

2 **Breadfruit** (*Artocarpus communis*) differs from most 'fruits' in that its main nutritional constituent is starch; it is usually eaten roasted, and forms a staple part of the diet in some Pacific islands to which it is native. The fruits, up to 8 inches in diameter with a thick warty rind, are formed from the whole female inflorescence; they are borne in twos or threes at the end of the branches of a large tree which may grow up to about 90 feet high. The best varieties are seedless, though in seeded varieties the seeds as well as the pulp may be cooked and eaten. Trees are propagated by root-suckers or by seed. The breadfruit is associated with the famous historical incident, the mutiny of the '*Bounty*' in 1787, for that voyage of Captain Bligh's was commissioned by the British government to introduce breadfruit plants from Tahiti to the West Indies. Although breadfruit trees have now been established in most parts of the tropics, the fruit has not found much popularity outside its original home in the Pacific and south-east Asia.

Jak or **Jack Fruit** (*Artocarpus integrifolia*) is a related species with enormous fruits which can weigh up to 70 lb. each. In spite of their very strong odour, they are relished especially in Asia and may be eaten cooked or raw.

3 **Avocado Pear** (*Persea americana*) also has unusual qualities in that it contains more protein than any other fruit and up to 25 per cent of fat; these properties put it in the class of a vegetable as much as a fruit. The plant, which is propagated by seed or budding, is a small to medium-sized tree. Native to Central America, it has now been widely planted in the tropics and a few subtropical areas such as Florida and California where some superior varieties have been selected. The fruits are somewhat the size and shape of a pear; in different varieties the ripe fruit remains green or takes on yellowish or crimson tints.

To eat, the fruit is split open and the inedible stone in the centre removed. The flesh is then scooped out and eaten raw, a sophisticated taste being to add a dash of Worcester sauce. The pulp can also be diced and used in either vegetable or fruit salads. The fatty pulp can also be spread like butter on bread to make sandwiches. The avocado pear is a particularly valuable fruit in tropical diets because of its high content of protein and of fat and of the vitamin B complex and the fat-soluble vitamin A. The fruit, being easily damaged by bruising, is difficult to transport without losses, but small quantities appear in the markets of temperate countries.

FRUITS AND FLOWER DETAILS × ½   PLANT AND BRANCHES × 1/18

1 PAWPAW fruit   1A Plant   1B Female flower detail   1C Male inflorescence and flower detail
2 BREAD FRUIT   2A Male and female flowering branches   2B Detail of female inflorescence
3 AVOCADO PEAR   3A Fruiting branch   3B Detail of flowers

CUCUMBERS AND GHERKINS

Cucumber (*Cucumis sativus*) is believed to have originated in southern Asia. The evidence is inconclusive, but the plant is known to have been cultivated in India for a very long time and to have spread to the west where it was appreciated by the Ancient Greeks and the Romans. Cucumbers are commonly eaten raw, in salads and sandwiches, or pickled. They can also be used grated in soup, or fried, or boiled and served in a white sauce. Their food value lies chiefly in their sugar and vitamin content.

Cucumis sativus (family Cucurbitaceae) is a roughly hairy, trailing or climbing plant, climbing by means of unbranched tendrils borne in the axils of the alternate, triangular-ovate, 3-angled or 3-lobed, irregularly toothed leaves, which are 3 to 6 inches or more in length. The short-stalked, yellow flowers, about 1 inch across, are also borne in the axils of the leaves. There are often several male flowers (1A) per axil, opening successively. The female flowers (1B), usually solitary or in pairs, are recognisable at an early stage by the appearance of their inferior ovary — the young cucumber fruit — at the base of the flower.

Horticulturally, cucumbers may be divided into two main groups: indoor or long (2); and outdoor or ridge (5) which are generally much shorter.

The indoor cucumbers usually require artificial heat and high humidity. These special conditions may be unsuitable for other greenhouse crops, such as tomatoes, so commercial growers normally grow them in separate cucumber houses. 'Butcher's Disease Resister', 'Rochford's Market', and 'Improved Telegraph' (2) are the principal commercial varieties in Britain. 'Conqueror' is less demanding in its temperature and humidity requirements. It can be grown in cold houses and is the best variety for amateur gardeners who wish to grow both cucumbers and tomatoes or other crops in one greenhouse. Indoor cucumbers are parthenocarpic, that is, the fruits are produced without fertilization. In fact, it is necessary to prevent pollination, either by removing the male flowers as soon as they appear, or by growing the crop in bee-proof greenhouses, as bees and large flies are often the pollinating agents. Fertilized indoor cucumbers become swollen at the end and often taste bitter. Indoor cucumbers are at their best when 12 to 15 inches long.

Ridge cucumbers (5 – 6) are not parthenocarpic; fertilization is necessary. Their fruits are usually smaller than indoor cucumbers, but they are often produced prolifically. Planting on ridges is not necessary, in fact ridge cucumbers are usually grown commercially as an ordinary field crop.

In recent years, Japanese outdoor cucumbers have aroused interest in many countries and several cultivars are available from British seed merchants. They are generally larger than the average ridge cucumber and compare favourably in yield. Examples are 'Kaga' (3), 'Suyo' (4), and 'Kariha'. The 'Apple Cucumber' is a distinctive variety which produces roundish fruits, up to the size of an orange.

Gherkin (*Cucumis anguria*) (7). This is a weed of damp places in the West Indies and in tropical and sub-tropical America, grown for pickling mainly in the U.S.A. It has deeply 5-lobed leaves and a small, prickly fruit, 1 to 3 inches long, ovoid at maturity. It is immature fruits that are used for pickling. They are soaked in brine and treated with boiling vinegar. Small ridge cucumbers, pickled, are sometimes sold as 'gherkins', for example, the variety 'Venlo Pickling' (6).

PLANT × ⅛   FRUITS AND FLOWERS × ½

1 CUCUMBER plant   1A Male flowers   1B Female flowers
2 'IMPROVED TELEGRAPH'   3 'KAGA'   4 'SUYO'
5 RIDGE CUCUMBER   6 'VENLO PICKLING'   7 GHERKIN

MELONS

Melons belong to the same family as Cucumbers, Marrows, and Squashes (Cucurbitaceae). Their fruits are commonly eaten raw, either as a first course or for dessert, although some kinds are pickled or preserved, or used in soup. Their flesh consists of about 94 per cent water and, perhaps rather surprisingly, only about 5 per cent sugar (about half the sugar content of an apple or a pear). The seeds, stripped of their hard coats, may be eaten, and also yield an edible oil.

1 **Melon** (*Cucumis melo*) is an annual trailing herb which probably originated in tropical Africa. It is polymorphic, that is, has many forms, and it has given rise to a large number of cultivated varieties which have been developed to suit the varying climatic conditions in the warm temperate, sub-tropical, and tropical regions of the Old and New Worlds.

Melon plants have softly hairy, ridged stems, bearing 5- to 7-lobed leaves which are variably toothed. The flowers are yellow, the male flowers (1) occurring in small groups, the female flowers (1A) usually solitary. The latter are easily distinguished by the presence of the inferior ovary or young fruit at the base of the flower. Hermaphrodite flowers also occur regularly in some varieties. Melon fruits are very variable in size, in shape, in the colour, thickness, and smoothness or roughness of the rind, and in the colour of the flesh.

In warm or hot climates, melons are grown out of doors, in pits, on level ground, or on hills or ridges, depending on local conditions. The planting distance is about four or 5 feet apart, preferably in a rich loam. The plants take 3 or 4 months from planting to ripening their fruits.

In Britain, melons can only be grown successfully under glass and artificial heat is needed, at least for germination of the seed (about 65°F.). Some of the hardier varieties can be grown in cold frames, others will not thrive unless a night temperature of about 65°F. can be maintained in a heated greenhouse. British hothouse melons may be grown throughout the year. The planting distance is usually 18 to 24 inches apart, on a low ridge, and the plants are trained up wires. Hand pollination is usually considered to be necessary.

The different varieties, or groups of varieties, of *Cucumis melo*, overlap and interbreed so readily that attempts to find a precise botanical classification can never be completely successful. However, several broadly defined groups may be recognised, according to the form, colour, and smoothness of their fruits.

2 **'Winter Melons'** are either smooth or shallowly corrugated, but not netted. Their flesh is not strongly scented. They ripen late, are hard-skinned, and can be stored for a month or more. Consequently, they are popular with growers in warm countries who export to distant markets. Here belongs 'Honey Dew', a smooth, white-skinned variety with pale green flesh, and those dark-green, shallowly corrugated melons imported into this country from Spain in large quantities and often erroneously called 'Honey Dew'.

3 **'Musk Melons'**, also called 'Netted Melons' or 'Nutmeg Melons', are usually distinctly netted, with a raised network generally lighter than the overall colour of the fruit, which may be yellowish or green. The surface of the fruit may be smooth (apart from the network) or segmented into broad ribs and grooves. The aromatic flesh is green to salmon-orange. Most British hothouse melons belong to this group.

4 **'Cantaloupe Melons'** have a warty or scaly rind, but are not netted. They are often deeply grooved and usually have orange coloured, rarely green, aromatic flesh. Cantaloupes are commonly grown in Europe and a few varieties are hardy enough to be grown in Britain, with the protection of cold-frames or cloches.

5 **'Ogen Melon'.** The name is derived from a kibbutz in Israel where it was bred and whence it has been exported during the last decade. It is said to belong to the cantaloupe group of varieties. The fruit is relatively small — about 6 inches across — bright orange-yellow ribbed with green, and with sweet, green, aromatic flesh.

HALF LIFE SIZE

1 MELON male flowers   1A Female flower   2 'WINTER MELONS'
3 'MUSK MELONS'   **4** 'CANTALOUPE MELON'   **5** 'OGEN MELON'

WATER MELON AND GOURDS

1 Water Melon (*Citrullus vulgaris*) is a native of tropical Africa, with probably a secondary centre of diversification in India. It has been cultivated for many centuries and is now widely grown in warm-temperate, subtropical, and tropical regions all over the world. In Britain it needs glass-house treatment and is rarely grown. The plant is an annual climber, with more or less hairy leaves which are often deeply 3- or 5-lobed, with the lobes themselves often pinnately lobed. The yellow corolla of the flower is about 1½ inches across. The fruit is usually ellipsoidal, often 10 inches across, with whitish, yellow or red flesh, sweet and juicy but rather insipid. It is picked only when fully ripe. In some regions, the seeds, which are variable in colour, are eaten. They are oily and nutritious.

2 Balsam Pear or **Bitter Gourd** (*Momordica charantia*). This is a tropical climber of the family Cucurbitaceae with an elongated fruit up to 8 inches in length, usually orange-yellow in colour though there is a white variety. It is cultivated in India and eastern Asia. The young fruits are cooked and eaten; they are often steeped in salt water after peeling and before cooking to remove the bitter taste. Pieces of the fruit are a common ingredient in Indian pickles and sometimes used in curries. The tender shoots and leaves can be cooked as a kind of spinach.

3 Snake Gourd (*Trichosanthes cucumerina*) is another tropical gourd grown for food. It is generally given supports to climb up and is an annual which begins to bear fruit 3 – 4 months after sowing the seed. The fruits are narrow and cylindrical and can be up to 6 feet in length; a weight is often attached to their ends to keep them growing straight. They are picked when immature, and sliced before boiling. The plant grows wild in Asia and Australia and is mostly cultivated in India and the Far East, more rarely in Africa.

Other Tropical Cucurbitaceae. There are a number of other members of this family, not mentioned elsewhere in this book, which are used for food in the tropics. Certain gourds which are chiefly grown for other purposes are occasionally used as food. One of these is the bottle gourd, *Lagenaria siceraria*, which is mainly grown for the dry hard shells of the fruits which are used as containers; the young fruits of its less bitter varieties are sometimes boiled as a vegetable. The same is true of the gourd *Luffa cylindrica*, whose chief use is to provide loofahs. The wax or ash gourd (*Benincasa hispida*) is grown, chiefly in Asia, for the use of its fruits as a vegetable.
Cucumeropsis edulis and *C. manii* are the 'egusi' melons of West Africa, cultivated solely for their oily seeds which are cooked and eaten (*see* text figure).

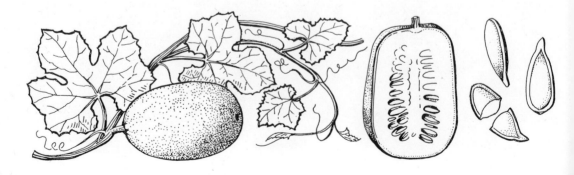

EGUSI MELON Fruits and Vine (× ¼) Seeds (life size)

HALF LIFE SIZE

1 WATER MELON   1A Female flower
2 BALSAM PEAR   3 SNAKE GOURD

MARROWS, SQUASHES, AND PUMPKINS

These members of the family Cucurbitaceae are eaten as cooked vegetables. They are like their relatives, cucumbers and melons, in being all trailing or climbing herbs, with tendrils and with large, often palmately lobed leaves. Their flowers are unisexual and generally yellow. Their fruit is a pepo (a fleshy, berry-like structure).

1-2 Vegetable Marrows, as well as many pumpkins (4), and summer squashes (3) are varieties of a single species of uncertain but probably American origin, *Cucurbita pepo*. Their stems and leaves are roughly hairy and harsh or scratchy to the touch. The leaves are mostly deeply lobed, toothed and triangular-tipped. The short fruit-stalk is deeply furrowed, with 5 to 8 ridges and is only slightly swollen where it joins the fruit.

Vegetable marrows are usually classified in seed catalogues as either 'trailing' or 'bush' varieties. The former produces stems which are often several yards long; the latter do not form a true 'bush' but they do have a more compact habit of growth. Vegetable marrows may be green, whitish, or irregularly striped green and white. The varieties most popular in the British Isles are those which produce large, oval-cylindrical fruits, several pounds in weight when mature. Marrows are often eaten as a boiled vegetable, but they make a tastier dish if parboiled and then stuffed with savoury mixtures of fried minced meat, or sausage meat, onions and tomatoes, etc. before being baked in the oven until tender.

3 'Courgette' (5). This French marrow variety has, like the Italian 'Zucchini', been developed for cutting when only a few inches long, when they may be cooked whole, in various ways — fried, with garlic and tomatoes, or steamed and served with mornay sauce. When slightly older (up to about 9 inches long), they can be sliced into rings, about ½ inch thick, and cooked in the same ways, but when mature they are no different from British marrows.

4 'Custard Marrow' or **'Scalloped Summer Squash'** has probably been grown in Britain for about 400 years, but has never achieved much popularity. Its fruits have the shape of a thick round cushion, with scalloped edges. The varieties most readily available are whitish or yellow in colour.

5 'Summer Squashes' are much more popular and available in greater variety in North America than they are in Britain. One of the best known varieties is 'Summer Crookneck' (3) which has bright yellow or orange, warty fruits, shaped like a crooked club. The terms 'summer squash' and 'winter squash' are a little misleading, as some of the former (including 'Summer Crookneck') can be kept for winter use, although they are best eaten when young.

6 Pumpkins as grown in the British Isles, are usually large, round and yellow. They are eaten when mature, usually as 'pumpkin pie', which may be either a savoury or a dessert. Pumpkin seeds are rich in both fats and proteins, and an edible vegetable oil is derived from them. *Pepitos* are seeds which have been fried in deep oil and salted.

The Cushaw pumpkin belongs to a different species, *Cucurbita mixta*. It is popular in America but takes too long to mature (140 – 150 days) to be successful in the British climate.

7 'Winter Squashes' (*Cucurbita maxima*) have foliage which is hairy but softer to touch than *Cucurbita pepo* varieties. The leaves have a deeply cordate base and are only shallowly lobed. The fruit stalk is swollen and is not prominently ridged. Here belong the 'Hubbard' (2) and 'Turban' squashes. They are very popular in America, but as they are eaten when mature and take about 110 days to reach this stage (twice as long as vegetable marrows, which are eaten when immature), they are less reliable for sowing in the open ground in the British Isles and are normally sown in pots under glass and planted out when all danger of frost is past. Winter squashes are cut in the autumn and can be kept for 3 or 4 months or longer in a frost-free store. The texture is firmer and more floury than that of a marrow.

Winter squashes contain less water, more protein, fat, carbohydrates, and considerably more vitamin A, than either summer squashes and marrows or cooked cabbage.

Jams, chutneys, soups, and home-made wines are some other ways in which marrows and squashes can be used. The flowers can also be eaten, stuffed with a savoury mixture, then dipped in batter and fried.

8 Chayote (*Sechium edule*). A native of Central America, the chayote is also grown in the West Indies and other warm regions. The fruits, young shoots, leaves, and large fleshy roots are all used as culinary vegetables. The fruits in particular have many uses — creamed, baked, fried, in sauces, puddings, tarts, and salads. There are several variants, differing in the shape, size and colour (from whitish to dark green) of their fruits, but they are not clearly separated from each other.

The Chayote is a vigorous, herbaceous, perennial vine, with heart-shaped, angled or lobed leaves, 4 to 6 inches across. The greenish or cream-coloured flowers are axillary; the male in small clusters, the female solitary. The fruit is more or less pear-shaped or roundish, 3 to 8 inches long, with a single large seed 1 to 2 inches long. The large, tuberous root, up to 20 lb. in weight, looks and tastes like a yam (*see* p. 183).

FRUITS AND LEAVES × ⅛   FLOWER DETAILS × ½

1 MARROW male flower   1A Detail   1B Detail of female flower

2 VEGETABLE MARROW   2A Part of plant

3 COURGETTE   4 CUSTARD MARROW   5 SQUASH 'SUMMER CROOKNECK'

6 PUMPKIN   7 SQUASH 'HUBBARD TRUE'   8 CHAYOTE

TOMATOES

This native of the lower Andes in South America is now widely grown throughout the world for the fruit, valued for its high vitamin content and for the variety of ways in which it can be served.

1-2 Tomato (*Lycopersicon esculentum*). The tomato was first introduced into Europe via Italy in the 16th century, and soon spread into other European countries and to most parts of the world. When first introduced, it was known as the 'Golden Apple' (*Pomo d'oro*), 'Love Apple', or 'Peruvian Apple'. It is now a familiar ingredient of salads, its attractive colour being appreciated as well as its flavour. It is eaten, fried, baked or stuffed, and adds both flavour and colour to soups, sauces and ketchups. Large quantities are canned, either whole or as tomato juice or purée.

Selective breeding of the tomato plant has produced many cultivars which differ from the original introduction, and from each other, in size, shape, colour, uniformity, habit of growth, productivity, hardiness, and resistance to disease. In warm climates tomatoes are grown out of doors. In Britain and most countries with a similar climate, however, most tomatoes are grown under glass, commercially. With the aid of artificial heat, production can be extended over the greater part of the year. A night temperature as high as 55° – 60°F. is needed for satisfactory results. Artificial illumination with mercury-vapour lamps is advantageous during the early stages of growth, when the natural day-length is short.

The tomato is a weak-stemmed herbaceous plant, capable of perennial growth but normally cultivated as an annual. Under natural conditions, it forms a spreading, straggling bush; some modern cultivars, such as 'Amateur', can be allowed to grow without support and with a minimum of pruning. The fruits of 'bush' tomatoes need protecting from contact with the soil by means of a layer of straw. Most cultivated tomatoes are trained as a single stem up a string or cane and all side shoots are removed soon after they appear. In glasshouses, the stem is trained up a string tied to an overhead wire, the string being twisted around the stem as the plant grows. The tip of the stem is removed when it reaches the top of its support or has produced the required number of fruit trusses. As tomatoes are vulnerable to potato blight, it is advisable to spray each plant thoroughly at intervals of about 2 weeks with a fungicide, such as Bordeaux mixture, when blight conditions are forecast, usually about the beginning of July. Outdoors, the fruit-trusses formed after the end of July are unlikely to ripen, and are usually removed.

The leaves of tomato plants (1B) are alternate, pinnate, or bipinnate, with their segments variously lobed or toothed. All the green parts of the plant bear golden yellow glands and give off a characteristic odour when touched. The green parts contain poisonous alkaloids (solanines) and have been known to cause the death of livestock. The flowers, borne in racemes of 3 to 11 or more, have a 5-lobed, green calyx and 5 yellow petals (1A). In modern cultivars, the anthers form a cone enclosing the style, thus ensuring self-pollination — except when hybrids are deliberately produced by transferring the pollen from one flower on to the style of a flower on another plant. For pollination, gentle shaking of the flower trusses is all that is needed and in greenhouses this is usually effected by spraying them with water. The tomato fruit is a fleshy, juicy berry, usually red (1) but yellow in a few cultivars (2). Uniformly shaped tomatoes, sub-globose, with a hollow at the calyx end, are most popular in this country, but tomatoes can also be egg-shaped, or irregularly globose with bulges and ridges. Some of the latter may have individual fruits up to 1 lb. in weight. The tomato fruit consists largely of water but it has quite a high sugar content and is also a useful source of vitamins A and C. Variation in the acidity of different tomatoes has a marked influence on their flavour.

3-4 'Cherry Tomato' (var. *cerasiforme*) (3). This variety is characterised by its smaller and more numerous fruits. It is grown primarily as an ornamental curiosity, but its fruits are palatable and make an interesting addition to salads. The same is true of the 'Pear Tomato' (var. *pyriforme*) (4) which has pear-shaped fruits about 1½ inches long. Both cherry and pear tomatoes may be either red or yellow in colour.

5 Tree Tomato (*Cyphomandra betacea*). This belongs to the same family as the tomato, the Solanaceae. It is a small short-lived tree, propagated by seed or cuttings, and begins bearing in its second year. Native to Peru, the tree tomato is a tropical fruit of minor importance which does best at medium to high altitudes. The fruits are the size of an egg, reddish-yellow or in some varieties purple when ripe; they can be eaten raw, but are usually stewed.

FRUITS AND FLOWERS × ⅔   *FRUITING BRANCHES* × ¼

1 TOMATO   1ᴀ Flowers   1ʙ Fruiting branch
2 'GOLDEN TOMATO'   3 'CHERRY TOMATO'   4 'PEAR TOMATO '
5 TREE TOMATO   5ᴀ Flowers   5ʙ Fruiting branch

125

SOME PLANTS OF THE POTATO FAMILY,
WITH EDIBLE FRUITS

1 Garden Huckleberry (*Solanum intrusum*). Said to be a native of Africa, but of uncertain natural distribution, this plant is sometimes grown for its fruits in America and, more recently, in Britain. The fruit is used in pies and preserves, but is rather insipid and has not gained much popularity. Apparently it is harmless, yet the fruits of the very similar black nightshade (*Solanum nigrum*) may contain solanine alkaloids in sufficient quantity to be poisonous. The garden huckleberry was at one time called *S. nigrum* var. *guineense*. It differs from *S. nigrum* in having larger, ovate, entire leaves (1B), brownish yellow anthers (1A), and larger fruits.

2 Aubergine (*Solanum melongena*) is a native of tropical Asia, widely grown in tropical, subtropical, and warm temperate regions for its fruit, which is eaten as a cooked vegetable, sliced and fried, or broiled, or eaten in curries and other dishes. It is known also as 'egg-plant' because the fruit of many varieties looks something like a large egg, and as 'brinjal' or 'bringall', a name used in India and adopted by the British in colonial times.

The aubergine is a perennial, usually grown as an annual, with erect or spreading, tough, herbaceous, branching stems (2B), 2 to 8 feet high. Its ovate, wavy-edged or lobed leaves are 3 to 6 inches long, felted with stellate hairs beneath and bearing prickles which are more numerous on wild than on cultivated plants. The flowers (2A), borne in few-flowered, lateral cymes, have a deeply lobed and toothed calyx, usually bearing a few prickles and a purplish corolla, about 1 to 1¼ inches across. The fruit (2) is a glossy firm-fleshed berry, containing numerous seeds. It may be egg-shaped, oblong, or sausage-shaped, from 4 to 12 inches long, and either white or deep purple in colour, in different varieties.

The aubergine can be grown under glass in this country, either in large pots or in the greenhouse border. It is not an important commercial crop in Britain, because the demand is limited and the grower has to compete with imported fruit from continental countries where the plant can be grown out of doors at less cost.

3 Ground Cherry (*Physalis pruinosa*). This is one of several species of *Physalis* which are collected from the wild in their native countries and are occasionally cultivated for their fruits. Although the fruit may be eaten raw, it is more usually boiled and used in stews, sauces, and preserves.

The ground cherry, which is also called 'Strawberry Tomato' and 'Dwarf Cape Gooseberry', is an annual, native in parts of eastern and central North America. Its spreading branches and heart-shaped, shallowly-toothed leaves (3B) are greyish green when young with a dense covering of soft hairs. The leaves are often unequal at the base, with one side larger than the other. The flowers (3A) are about ⅜ inch long, with 5 brownish patches in the throat. The fruit (3) is a roundish, yellow berry, ¾ inch across, pleasantly sweet and slightly acid, enclosed within the lantern-like, light brown calyx or husk.

4 Cape Gooseberry (*Physalis peruviana*). The fruit is usually less sweet than that of the previous species but is used for similar purposes. The plant has been cultivated outside its native South America for about 200 years. By the beginning of the 19th century it had become one of the most important cultivated fruits in the Cape of Good Hope — hence one of its names. The plant is a perennial, similar in appearance to *Physalis pruinosa*, but with leaves (4A) which are not unequal at the base, and flowers which are ½ inch or more in length and are often a brighter yellow. The fruiting calyx or husk is considerably thicker and larger (4).

The Bladder Cherry or Chinese Lantern Plant (*Physalis alkekengi*) is grown mainly for the ornamental value of its large, red, fruiting calyx. The red berry is edible, but the calyx should not be eaten.

5 Tomatillo or **Jamberry** (*Physalis ixocarpa*). The large, sticky berry is used mainly in sauces and preserves, being too flavourless to be worth eating raw. The plant is a native of Mexico and is cultivated elsewhere. It is a perennial, but often grown as an annual, less hairy than the previous species and bearing generally smaller leaves which are variously toothed (5B). The flowers are bright yellow and about ¾ inch across (5A). The fruit is larger, purplish in colour, and fills its yellowish husk completely. 'New Sugar Giant' (5), which is included in British seedsmen's lists under this species, is probably of hybrid origin. Its leaves are not toothed, and the fruit is yellow, not purple, in colour.

FRUITS AND FLOWER DETAILS LIFE SIZE   BRANCHES × $\frac{1}{4}$

1 GARDEN HUCKLEBERRY   1A Flowers   1B Branch   2 AUBERGINE   2A Flower   2B Branch
3 GROUND CHERRY   3A Flower   3B Branch   4 CAPE GOOSEBERRY   4A Branch
5 JAMBERRY 'NEW SUGAR GIANT'   5A Flower   5B Branch

PEPPERS AND CHILLIES

The word 'pepper' is used in English to describe a group of spices which are derived from two quite different kinds of plants. 'Red' or 'Cayenne' pepper is obtained from bushy plants of the genus *Capsicum*, and 'white' and 'black' pepper from the climbing vine *Piper nigrum* (4).

The CAPSICUMS (1 – 2 – 3) are a group of plants, native to tropical America and the West Indies, with a very wide range of varieties and many intermediate and hybrid forms. Botanical authors have classified them in different ways, and many different common names have been applied to them. The generally accepted two main species are described below.

1- Sweet Pepper (*Capsicum annuum*). This is an annual
1A species, sown from seed; the plants begin bearing at about 2½ months, and harvesting may continue for 3 months. The species is of tropical American origin but has now been spread all over the tropics, where it will grow from sea-level up to altitudes of 6,000 feet or more. Although killed by frost, it can be grown outside the tropics where summers are hot enough, and is widely planted in southern Europe and the southern United States. The flowers are borne singly in the leaf axils. When ripe, the fruits are red, yellow or brown; but immature fruits of the large mild kinds are often picked while still green, for use in salads and other dishes. This species includes all the larger-fruited kinds, with some fruits up to 10 inches long, but many varieties have much smaller ones. The fruits also vary greatly in shape, from long and narrow to almost spherical. Figure 1 shows immature green and ripe red fruit from the variety 'Rising Sun', and Figure 1A a ripe fruit of another variety. Varieties also differ greatly in their pungency, which is due to a chemical compound in the fruit, called 'capsicin'; some, especially of the larger-fruited kinds, are quite mild in taste and are known as 'sweet peppers', while others are hotter. In general, the term 'paprika' is applied to European types with large mild fruits; Spanish paprikas are called 'pimiento'. Some of the larger fruits are called 'bell pepper' and 'bullnose pepper'. The fruits of *Capsicum annuum* are eaten as a vegetable, in salads and in pickles; paprika is used in making Hungarian goulash. These fruits have a very high content of Vitamin C.

2-3 Red Pepper and **Chilli** (*Capsicum frutescens*) is a perennial plant. It is often grown as an annual in the tropics, but having a longer growing season than *C. annuum* and not being frost-hardy, is more strictly confined to the tropics. In this species, the flowers are borne in clusters of two or more in the leaf axils. The fruits are in general much smaller than those of *C. annuum*, bright red in colour, and with a wide range of shape and size. Different types are illustrated in Figures 2 and 3. Pungency is again variable, but in general greater than in *C. annuum*, and some fruits are very hot indeed (causing irritation to the hands of the pickers). The species in general, and especially the hotter kinds, are often called 'chillies' (a word which has several variant spellings). Some of the smallest-fruited kinds are called 'bird' or 'bird's eye' chillies. This species is the typical *Capsicum* grown by small farmers for their own use in most tropical countries. The fruits are usually sun-dried, which gives them a wrinkled appearance (3B). The tropical housewife simply shreds them to lend flavour to the many rather tasteless foods eaten in the tropics. Chillies are also an essential ingredient in curry powder and are used in pickles and in making tabasco sauce (by pickling the pulp in brine or vinegar). Cayenne pepper is made from the powdered dried fruits. The content of vitamin C is less than in *C. annuum* but still useful. The country exporting most Capsicum fruits is India, followed by Thailand; East and West Africa and Spain are also important exporters. The biggest importers are Ceylon, the United States, and Malaysia.

4 Pepper (*Piper nigrum*) is a native of India, still chiefly grown in southern Asia and requiring a wet tropical climate. The plant is propagated by cuttings. Being a climbing vine, it needs the support of living trees, stakes, or a trellis. Harvesting of the small fruits, which are borne in long hanging spikes (4A), usually begins in the third year and may continue for 15 years or more. The fruits, which are called peppercorns, turn red when ripe. Both white and black pepper are made from them. For black pepper, unripe pepper-corns are simply sun-dried, during which process the outer skin of the fruit turns black and wrinkled (4B). For white pepper, ripe peppercorns are soaked and the outer coverings rubbed off the seed (4C). The pepper-corns are marketed at this stage and later ground to the familiar powder.

Pepper has been historically the most important spice in world trade. In modern times, exports sometimes reach 75,000 tons in a year. Pepper was originally used in India largely in curry powders, in which it was the most pungent ingredient until chillies were introduced from America in the sixteenth century. Pepper is now used to flavour all kinds of savoury foods in most countries of the world. India is the largest exporter, followed by Indonesia and Sarawak. The biggest importers are the United States (which takes about a third of the world supply), Germany, and Britain.

Japan Pepper (*Zanthoxylum piperitum*). The black seeds are ground and used as a condiment in China and Japan. They have a pungent, pepper-like flavour. Japan Pepper is a deciduous shrub or small tree, hardy in the British Isles. Its downy shoots bear pairs of flattish spines, ½ inch long, and pinnate leaves with up to 23 oval leaflets. The small green flowers are succeeded by round, reddish fruits, dotted with glands.

SPICES AND FLAVOURINGS (1)

1 Vanilla (*Vanilla planifolia*). The plant is a climbing orchid native to Central America and now grown in a number of tropical countries with high rainfall, though a dry season is desirable for the best ripening of the pods. The plants are propagated by stem cuttings, and support for them is provided by a trellis or by planting small trees of various kinds up which they can grow. Aerial roots which arise from the stem attach the plant to its support. In order to ensure a good set of fruit, artificial self-pollination is often carried out by inserting a small pointed stick into the flower; the pollen-masses adhere to the stick and can then be transferred to the stigmatic surface. Bearing begins about 2½ years after planting. The pods have to be very carefully cured to prepare them for market; this may involve dipping them in boiling water followed by a long slow drying process lasting about six weeks. When properly cured, the pods should be black in colour, as in Figure 1A, and show visible crystals of vanillin on their surface. They are then packed, in tins or in boxes lined with tinfoil, for export. The major producing country is Madagascar, but vanilla is also exported from many others, including Seychelles, Réunion, Mexico, and Dominica in the West Indies.

Vanilla is used in flavouring all kinds of confectionery, usually in the form of vanilla essence. Its main flavouring component, vanillin, can now be made synthetically, but natural vanilla remains in demand because it contains other useful aromatic substances as well as vanillin.

2 Nutmeg and **Mace.** These are two spices which are products of the same tree, *Myristica fragrans*. Growing 60 feet or more high, this tree is a native of the Molucca Islands in Indonesia, and was not taken to other parts of the tropics until late in the eighteenth century. Most trees are either male or female, and one male is planted to about ten female trees; planting is normally from seed, but propagation by cuttings is being experimented with. The fleshy fruit splits when ripe and exposes the mace as a scarlet fleshy covering called an aril (2A) enclosing in a network the seed, which is the nutmeg (2B). The mace is prepared for market by being carefully removed and flattened, and then dried, during which process it turns a browner colour as shown in Figure 2C. The nutmeg is also dried before being marketed. The only producing countries of consequence are Indonesia and the island of Grenada in the West Indies.

Mace is used in the preparation of savoury dishes, and especially for flavouring sauces and ketchups. Nutmeg is generally sold in powdered form, though the whole seeds can also be bought for grating at home; it is used mostly for flavouring sweet dishes, and especially those made with milk.

3 Cinnamon (*Cinnamomum zeylanicum*) is in the wild state a tree up to 60 feet high which grows in south India and Ceylon. In cultivation it is planted from seed, and the first harvest of bark, which is the desired product, is taken after about two years. The young shoots are cut off close to the ground, and the bark, which peels easily if cutting is done in the rainy season, is removed in two long strips. The outer skin of the bark is scraped off, and the strips slowly dried into what are called 'quills'. These are of a pale brown colour, curling into a semi-tubular shape as they dry, and can be folded inside each other for packing (3A). The trees produce the next crop of shoots, which may be ready for cutting in another two years, and continuous harvesting in this manner produces a dense bushy plant not more than about 7 feet tall and never allowed to attain its natural height.

Cinnamon as a spice is used in curry powders and for flavouring confectionery. Cinnamon oil, which is used in medicinal products, is distilled from waste or broken bark, and leaf oil may be separately distilled. Ceylon is the chief cinnamon producing country; cinnamon oil is also exported from the Seychelles.

4 Cardamoms (*Elettaria cardamomum*). These are the fruits of a perennial herbaceous plant native to India and Ceylon. The crop is planted from seed or division of the plants, preferably under a light natural shade, and from the second or third year onwards the harvesting of the capsules is carried on. It is the seeds which contain the delicate spicy essence but they retain it much better if they remain enclosed in the capsule (4A). The capsules are therefore harvested very carefully; they are cut off with scissors before they are fully ripe, and are then dried slowly with the object of preventing them from splitting.

Cardamoms are grown in India and Ceylon and most of the production is consumed in those countries. They are now grown in Central America too, especially in Guatemala. They are chiefly used in preparing curry powders but are also valued for chewing and for use in confectionery. They provide part of the flavour in certain liqueurs.

FLOWERING AND LEAFY SHOOTS × ⅔   *FRUITS AND BARK* × 1

1 VANILLA ORCHID   1ᴀ Dried pods
2 MYRISTICA FRAGRANS flowers   2ᴀ Fruit   2ʙ Seed, NUTMEG   2ᴄ Aril, MACE
3 CINNAMON leaves   3ᴀ Bark   4 CARDAMOM flowers   4ᴀ Fruits

1 **Bay Laurel** (*Laurus nobilis*). This is an evergreen shrub or tree, up to 60 feet high, whose leaves are used for flavouring puddings, stews, and other dishes. The tree is a native of the Mediterranean region, and in classical times wreaths of laurel leaves were used for crowning the victorious. Bay Laurel gives its name to the Laurel family (Lauraceae), whereas the Cherry Laurel (*Prunus laurocerasus*), a more familiar garden shrub, is a member of the Rose family.

Bay leaves are very aromatic when crushed, dark green in colour, 2 to 4 inches long, and often wavy-edged. The inconspicuous, greenish-yellow flowers, opening in May, are succeeded by a glossy black berry, up to ½ inch across.

2 **Saffron** (*Crocus sativus*) has been used as a spice, a dye (saffron yellow), a cosmetic, and a medicine since classical times, but its commercial importance has decreased because the only part used is the orange-red, 3-branched style, which has to be picked by hand. Saffron is a member of the Iris family (Iridaceae). It is probably a native of Asia Minor, with a long history of cultivation. It bears flowers and leaves in autumn, and must not be confused with 'autumn crocus' (*Colchicum autumnale*), a poisonous member of the Lily family which has somewhat similar flowers, but does not produce its leaves until spring.

3 **Capers** (*Capparis spinosa*) are the unopened flower buds (3A) of a straggling, spiny shrub, native to the Mediterranean region. They are pickled and used in sauces.

The shrub gives its name to the Caper family (Capparidaceae). It bears roundish, alternate, rather thick leaves, each with a pair of spines at its base. The white or pinkish flowers are about 2 inches across.

4 **Black Mustard** (*Brassica nigra*) is a native of temperate Europe, long cultivated for its seeds, used with those of White Mustard (5) to produce mustard flour, from which various condiments are made — 'English mustard' with water, 'Continental mustards' with vinegar. Black Mustard is an annual herb, belonging to the Wallflower family (Cruciferae). It grows up to 3 feet high and has yellow flowers. Its fruits are cylindrical, ½ – ¾ inch long, with a prominent midrib on each of their two valves, and a slender, seedless beak. Its seeds are dark reddish-brown.

5 **White Mustard** (*Sinapis alba*) is used, with black mustard as a condiment, and also in the seedling stage, as a salad plant (*see* p. 153). In the fruiting stage, it is easily distinguished from Black Mustard by its bristly-hairy fruits, with broad, flattened beaks and by its yellowish or light brown seeds (5A).

6 **Cloves** (*Eugenia caryophyllus*) are the dried flower-buds (6A) of a tropical tree, up to 40 feet high, which is native to Indonesia. Clove exports were a Dutch monopoly until planting material was smuggled to other parts of the tropics in the eighteenth century. By the nineteenth century Zanzibar and Pemba had become the leading producers with Madagascar next. The trees are planted from seed, come into bearing at 8 – 9 years, and live for about 60 years, though many have been killed in Zanzibar by a fungus disease called 'sudden death'. The buds are picked by hand and dried in the sun. Ironically the greatest importer in modern times has been Indonesia, but small quantities of the spice are used in many countries for flavouring both meats and sweet dishes. Clove oil can be distilled from the buds, stalks, and leaves, and is used among other purposes for making vanillin, an artificial substitute for vanilla.

7 **Allspice** (*Pimenta dioica*) is a small tropical tree whose unripe dried berries (7A) provide the spice called allspice because it combines the flavour of several spices. It is also known as 'pimenta' and 'pimento'. Exports come mainly from Jamaica, where the tree grows wild and regenerates itself easily by seed.

Fenugreek (*Trigonella foenum-graecum*) (Fig. 8) is a native of the Mediterranean region. It is also cultivated in the sub-continent of India. Its seeds have been used medicinally since ancient times. In India the seeds are used in curries and the fresh plant is eaten as a vegetable. The ground seeds are used, with other aromatic ingredients, to produce an artificial maple flavouring, used in confectionery.

Fenugreek is an annual herb, belonging to the Pea family (Leguminosae) and growing about 2 feet high. Its white flowers are succeeded by slender, curved, prominently beaked pods, containing the oblong, or rhomboidal, diagonally furrowed, brownish seeds.

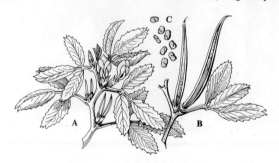

Fig. 8 FENUGREEK 8A Flowers 8B Fruits 8C Seeds

TUBEROUS ROOTED FLAVOURING PLANTS

1 Ginger (*Zingiber officinale*) is a tropical monocotyledonous plant which is propagated by division, seed being rarely formed. Good rainfall and fertile or well-manured soil are necessary for successful production. The plant is a native of tropical Asia where it has been widely cultivated since very early times. The parts used for food are the very irregularly shaped rhizomes (1A) which are formed at a shallow depth in the soil. These are dug up when the plants are about 10 months old. Preparation of good ginger involves a rather elaborate succession of washing, soaking, sometimes boiling, peeling and drying. These processes may be done in a factory, as they are in Australia, but much ginger is also prepared at home by tropical small farmers. Ginger is very widely grown in tropical countries for local consumption; exports come chiefly from West Africa, Jamaica and India. Preserved ginger, which is exported chiefly from China, is made from young fleshy rhizomes which are boiled with sugar and then packed in syrup. Although ginger rhizomes contain about 50 per cent of starch, they are chiefly valued for their pungent and aromatic qualities. Ginger is used in cookery to flavour many different foods in different lands, and is one of the important ingredients in curry powders. It is also used for making ginger beer.

2 Turmeric (*Curcuma longa*) also belongs to the ginger family, the Zingiberaceae, and has a similar habit of growth to ginger though the leaves are much broader. It is planted from pieces of rhizome, and the plants are dug up at about 9 months old. The rhizomes are prepared for use by washing, peeling, and drying; the product is used mainly in the powdered form. Turmeric rhizomes (2A) have a strong yellow colour due to the presence of a pigment called curcumin. (A dye prepared from them is widely used in India.) The rhizomes contain 30 – 40 per cent of starch but their main value derives from their content of pungent oil. Turmeric is one of the chief ingredients of curry powders, imparting a yellow colour as well as a spicy flavour to the curry. The crop is mainly grown for local consumption in India and other countries of tropical Asia. Only a small proportion of the crop is exported, and that almost entirely from India; the United States is the largest importer. Several wild species of turmeric grow in India and are sometimes collected for food.

3 Liquorice (*Glycyrrhiza glabra*) is grown for its dried rhizomes and roots, which are sometimes chewed but are more widely used in powdered form or as a liquid extract, for making sweets and soft drinks, cough pastilles, and cough mixtures. It has medicinal properties, as well as being a useful flavouring agent for disguising less pleasant ingredients. It was known to the ancient Greeks and has been grown in the British Isles since the 16th century, perhaps much earlier. Pontefract, in the West Riding of Yorkshire, was formerly a great centre of liquorice growing. Its cultivation in that area has almost died out, but 'Pontefract cakes' and other sweets are still made there from liquorice imported from Russia and the Mediterranean region.

Liquorice is a perennial herb belonging to the Pea family (Leguminosae), growing to a height of 3 or 4 feet. Its pinnate leaves are composed of 9 to 17 leaflets. The pale blue flowers, about $\frac{1}{2}$ inch long, are borne in a many-flowered raceme. They are followed by small pods, $\frac{1}{2}$ to 1 inch long, each containing 3 or 4 seeds. Under cultivation, the crop is allowed to grow for 3 to 5 years before being harvested, by which time it will have formed an extensive system of rhizomes and roots (3A), in well-drained soils reaching a depth of 3 or 4 feet and spreading for several yards. To produce liquorice extract, the rhizomes and roots are ground into pulp and boiled in water. The extract is then concentrated by evaporation.

4 Horse-radish (*Armoracia rusticana*) provides 'horse-radish sauce', eaten as a condiment with meat and made by crushing, mincing or powdering the root and simmering it with vinegar, milk and seasoning. It is pungent with a distinctive flavour, and off-white in colour.

Horse-radish is a member of the Wallflower family (Cruciferae). It is a perennial herb, forming a stout, yellowish-buff tap-root and bearing wavy-edged or lobed leaves about 18 inches long. It probably originated in south east Europe and western Asia, but as it has been cultivated since early times and has escaped from cultivation throughout Europe and elsewhere, its natural distribution is uncertain. It is said to have been one of the bitter herbs eaten by the Jews during the feast of the Passover.

PLANTS × ⅛   TUBERS AND ROOTS LIFE SIZE
1 GINGER   1A Tuber   2 TURMERIC   2A Tubers
3 LIQUORICE   3A Root
4 HORSE-RADISH   4A Root

PLANTS GROWN FOR MAKING OR FLAVOURING ALCOHOLIC DRINKS

1 Hop (*Humulus lupulus*) is a native of Europe (including the British Isles), and western Asia. It does not seem to have been widely used for brewing beer before the Middle Ages, although ale was flavoured with various bitter herbs in early times. Hopped ale was introduced from the Continent and later brewed in Britain from imported hops during the 15th century. Hops did not become a commercial crop in this country until about 1520. Today, they are grown in many parts of the world, where conditions are suitable, e.g. in Australia, New Zealand, and North America, as well as in Europe. The part used in brewing is the female 'cone', which consists of a cluster of pale, yellowish-green bracts and bracteoles, enclosing the small flowers and later the fruits. Resin glands, at the base of the bracteoles, produce glistening yellow spots of lupulin — the substance containing the essential oils and soft resins which give the hop its aroma and beer its flavour. The small male flowers (1A) are normally borne on separate plants and are quite different in appearance.

The hop is a perennial vine belonging to the Hemp family (Cannabiaceae). It dies down to near ground level at the end of the season and produces new shoots each spring. The shoots, or 'bines', grow rapidly, reaching a length of 18 to 25 feet (1B). In cultivation, they are trained up a framework of poles, wires and strings. Before World War II, hops were traditionally picked by hand. Now, about 90 per cent of the English crop is picked by machine.

The young shoots, which are thinned out in spring, are used as a boiled vegetable in some countries but have never become popular in Britain.

2 Juniper (*Juniperus communis*) is the source of Juniper berries, which are used for flavouring gin, liqueurs, and cordials. They have also been used medicinally since very early times for they contain an oil which has diuretic properties and which can be extracted by distillation. There are about 60 species of the genus *Juniperus*, belonging to the Cypress family (Cupressaceae), but this is the only one native to the British Isles. It also occurs in other parts of Europe, the Caucasus eastwards to the West Himalaya, and in North America. It is a shrub or small tree, with sharp-pointed, awl-shaped, evergreen leaves, which bear a broad, white band on their upper surface. The small, yellow male cones (2A) and the bluish-black female fruits are usually borne on separate plants.

3 Anise (*Pimpinella anisum*) has small, hard, greyish-brown, aromatic fruits (3A) which have been used for flavouring purposes since classical times. The liqueur 'Anisette' owes its name to them, but they are also used in other beverages, in soups, cakes, and sweets. Oil of anise, obtained from them by distillation, has medicinal properties and is used in cough medicines and lozenges. This oil also occurs, curiously enough, in a quite unrelated plant, the Star Anise, *Illicium verum*, a tree belonging to the Magnolia family, native to south-western China.

Pimpinella anisum belongs to the Hemlock family (Umbelliferae). It is an annual herb, probably a native of the Orient but long established in Europe and also introduced in Asia and North America. Usually about 18 inches high, it is a dainty-looking plant, with its finely divided and toothed leaves and compound umbels of small, white flowers.

4 Wormwood (*Artemisia absinthium*) is used for making the liqueur, absinthe, which can have harmful effects if taken habitually. It is also used in making Vermouth wine. In small doses, and if not taken habitually, the substances present in Wormwood have beneficial medicinal properties, as a stimulant bitter tonic.

Wormwood belongs to the Daisy family (Compositae) and is a native of temperate Europe (including the British Isles and Asia, northwards to Lapland and Siberia). It is a perennial herb, up to 3 feet high, with greyish-green, silky-hairy stems and leaves, the latter pinnately to tripinnately divided. The small, yellowish-green flower-heads are borne numerously on a much-branched terminal inflorescence.

Other Plants with Similar Uses. A number of other plants are locally used in different parts of the world for making or flavouring intoxicating beverages. Among the more widespread is kava (*Piper methysticum*), a bush native to some of the Pacific islands and now cultivated in most of them. After about 3 years' growth, the roots are dug and ground up or mashed with water. The liquid is fermented before use, but the active ingredient is an alkaloid which gives the drinker pleasant sensations. The bark of a wild South American tree, *Galipea officinalis*, is used in making Angostura bitters, a liquid which is industrially bottled in Trinidad and exported to many countries where it is used to flavour drinks (e.g. 'pink gin') and sometimes confectionery.

LIFE SIZE   DETAIL × 3   *BINES AND SHRUB SMALL SCALE*

1 HOP   1A Male flowers   1B Bines
2 JUNIPER   2A Male cones   2B Shrub
3 ANISE   3A Detail of seeds   4 WORMWOOD

137

AROMATIC UMBELLIFER SEEDS

1 **Caraway** (*Carum carvi*) is native from Europe to Siberia, northern Persia, and the Himalayas; it is doubtfully native in South-east England and occasionally naturalized elsewhere. This plant has been grown for its fruits, so-called 'seeds', since very early times. Caraway seeds are used in various parts of Europe for flavouring cakes, bread, cheese, soups, etc. They contain an aromatic oil which is a remedy for flatulence and which contributes its distinctive flavour to Kummel liqueur. The leaves have similar properties and have been used, when young, in soup.
Caraway is a much-branched, hollow-stemmed biennial herb, up to 2 feet high. Its bipinnate leaves have pinnatifid segments with deep, linear-lanceolate lobes. Its small, white flowers (1B) are borne in rather irregular umbels, often but not always subtended by a few linear bracts. The fruit (1A) is about ⅛ inch long, greyish-brown when ripe, and each carpel has 5 slender ridges.

2 **Coriander** (*Coriandrum sativum*). Probably a native of the eastern Mediterranean region, but widespread both as a cultivated plant and as a weed, coriander is occasionally found also as a casual in waste places in the British Isles. It is one of many plants reputed to have been introduced into this country by the Romans. The name *Coriandrum*, used by Pliny, is derived from the Greek word for a bug, referring to the foetid, bug-like odour given off by the fresh plant when bruised. The fruit loses this odour when dried and is used as a condiment, in curries, in some alcoholic beverages, and medicinally to remedy flatulence.
Coriander is a slender, solid-stemmed annual, about 2 feet high. The leaves are pinnate or bipinnate, the lower leaves divided into narrow, linear segments. The small flowers (2B) are white or pink; the middle ones in each umbel are infertile, the outer slightly larger and fertile. The globose fruits (2A), about ⅛ inch long, are prominently ridged and reddish-brown when ripe.

3 **Cumin** (*Cuminum cyminum*) is a native of the Mediterranean region, long cultivated in Europe and also in India and China. The fruit (3A) is used as a culinary spice, in curry powders, etc. It resembles caraway in odour and flavour but is rather bitter and less palatable. It was formerly used in herbal medicine as a stimulant and sedative.
Cumin is a small, slender branching, annual herb, about 6 inches high. The leaves are divided into a few thread-like segments, ½ to 2 inches long. The small, white or pink flowers are borne in few-flowered umbels, with thread-like bracts. The fruit (3A) is about ¼ inch long, narrowly oblong and bristly.

4 **Dill** (*Anethum graveolens*) is a native of Europe, but not of the British Isles, naturalized as far afield as America and the West Indies. It has been used for culinary and medicinal purposes since very early times. The fruits are used in pickling cucumbers, in dill vinegar, and for flavouring cakes and sauces. The young leaves are also used for similar purposes. Dill oil is a specific against flatulence and is used, diluted, as dill water for the treatment of 'wind' in infants.
Dill is a smooth-stemmed annual or biennial, about 2 feet high, with finely-divided, fennel-like leaves, their ultimate divisions up to ¾ inch long, basal sheath of larger leaves ½ to 1¼ inches long. The yellow flowers (4) are borne in umbels 2 to 6 inches across. The fruit (4A) is elliptic, slightly flattened, brownish with thin, yellow, dorsal ridges and narrowly winged lateral ridges, about ⅛ inch long.

5 **Fennel** (*Foeniculum vulgare*). A native of the Mediterranean region and possibly other parts of Europe, including sea cliffs in England, Wales and Ireland, fennel is naturalized in many temperate countries. The leaves are used in fish sauces and soups, and in salad dressings. The fruits cure flatulence and are included in the British Pharmaceutical Codex. They have an aromatic odour and flavour, resembling anise. Oil extracted from them is used in confectionery, condiments, pickles, cordials, and liqueurs.
Fennel resembles dill but is a perennial herb and generally taller, 2 to 4 feet high. It has larger leaves, with ultimate divisions ½ to 2 inches long, basal sheath 1¼ to 4 inches long. The fruit (5A) is oblong-ovoid, flattened, greenish or yellowish-brown or greyish, with yellow ridges, about ⅙ inch long.

LIFE SIZE   SEEDS AND FLOWER DETAILS × 3

1 CARAWAY   1A Seeds   1B Flower detail   2 CORIANDER   2A Seed   2B Flower detail

3 CUMIN   3A Seeds

4 DILL   4A Seeds   5 FENNEL   5A Seeds   5B Flower detail

AROMATIC LABIATE HERBS

The mint family, Labiateae, includes several more or less hardy herbs, used chiefly for flavouring meat and savoury dishes. Their flavour and aroma is due to the presence of aromatic essential oils, secreted by glands which can often be seen as translucent dots on the leaf, if it is examined under a hand lens.

1 **Peppermint** (*Mentha × piperita*). This mint is exceptional, in that it is not used for flavouring meat dishes but rather as the source of an essential oil, oil of peppermint, obtained by distillation from the fresh, flowering plants. It is used in cordials and various sweets. Mildly antiseptic and a remedy for flatulence ('wind'), it is often an ingredient of indigestion tablets and toothpastes.

Mentha × piperita is a hybrid between *Mentha aquatica* and *Mentha spicata*. It is a rather local plant, sometimes found in moist situations, on ditch banks, etc., in the British Isles. Peppermint is cultivated in England, France, Italy, Bulgaria, Morocco and America. It is recognisable as much by its characteristic odour as by its lanceolate, ovate, short-stalked leaves and its terminal, oblong spikes of flowers, their stamens more or less concealed within the reddish-lilac corolla.

2 **Spearmint** (*Mentha spicata*). This is a typically British herb, because of the national liking for mint sauce, usually made from this species, and served with lamb. Oil of spearmint, distilled like that of peppermint, has similar properties and uses, remedial and flavouring. Spearmint is a native of central Europe, but it is commonly cultivated throughout the British Isles and naturalized in moist waste places and roadsides. The lanceolate or oblong-lanceolate leaves are sessile or very shortly stalked and in the cylindrical and usually tapering spikes of flowers, the lower whorls are often more or less distant from each other. The stamens protrude well beyond the lilac corolla (2A).

Several other *Mentha* species and hybrids are occasionally grown in gardens.

3 **Sage** (*Salvia officinalis*). A native of Southern Europe and consequently best suited by dry warm regions, sage is grown commercially in parts of the British Midlands and the South and in gardens over a much wider area. The non-flowering form is propagated by cuttings or by division and renewed every 3 or 4 years; the flowering form is generally raised from seed and treated as an annual or as a biennial. Sage is sold, fresh and dried, in bunches, but the leaf is the only part used. The bulk of the crop is sold as packeted dried sage. It contains a pungent essential oil. Sage and onion is a traditional British stuffing for duck and goose.

Sage is a low-growing shrub, about 12 to 18 inches high. It has lanceolate to ovate, long-stalked leaves, sometimes lobed at the base, 2 to 6 inches long and either green and rather hairy, or greyish-green when densely hairy. The surface is wrinkled and the margin crenulate. The pink or bluish-lilac flowers are borne in distant whorls on erect inflorescences. The corolla is 2-lipped, the upper lip hood-like, the lower lobed. There are only 2 fertile stamens.

4 **Marjoram** (*Origanum* spp.). In British gardens and kitchens, the best known marjoram is *Origanum majorana*, Sweet or Knotted Marjoram, native in the Mediterranean region, perennial, but often unable to survive our winters and therefore usually grown as an annual. It is a bushy plant, 9 to 18 inches high, with stalked, elliptic, greyish hairy leaves, $\frac{1}{4}$ to 1 inch long. The purple or whitish flowers are borne in small, axillary clusters, usually in opposite pairs, along about $\frac{2}{3}$ of the stem.

Fresh leaves are gathered as required, or the whole plants are cut in summer, after flowering, and are dried and sold in bunches, like thyme and sage. The leaves are used, chopped, crushed, or powdered, for flavouring soups, stews, stuffings, pies, etc. They are a traditional component of the bunch of mixed herbs known as a 'bouquet garni'.

Pot Marjoram (*Origanum onites*) also comes from the Mediterranean region but is hardier. It is more robust, with sessile, ovate, hairy leaves, and whitish flowers borne in small, ovoid spikelets arranged in dense corymbs.

The only British native Marjoram is *Origanum vulgare* (4), a more herbaceous plant, 1 to 2 feet high, with stalked, ovate leaves, $\frac{1}{2}$ to 2 inches long. Its purplish-pink flowers (4A) are borne in small spikes, in numerous, rather flat-topped, corymbose or paniculate clusters, with purplish bracts. All the marjorams have similar culinary uses.

5 **Thyme** (*Thymus* spp.). The Common or Garden Thyme, *Thymus vulgaris*, is one of the most popular herbs for flavouring soups, stews, stuffings and sauces, either alone, or in a *bouquet garni*, or in packeted 'mixed herbs'. Although a Mediterranean plant, it is reasonably hardy in Britain, forming a small, bushy, sub-shrub, 6 to 18 inches high, thriving best in warm, light soils. It has small, greyish or green, aromatic leaves and small, pinkish, labiate flowers, borne in rounded or ovoid terminal clusters. Thyme is raised from seed, layers, or cuttings. It is harvested in summer and sold in bunches, green or dried.

6 **Lemon Thyme** (*Thymus citriodorus*) has the creeping habit of wild thyme, but its leaves have a very characteristic lemon-like scent. It is less frequently grown, and more difficult to harvest because of its habit.

LIFE SIZE   DETAILS × 4

1 PEPPERMINT   2 SPEARMINT   2A Flower detail   3 SAGE
4 MARJORAM   4A Flower detail
5 COMMON THYME   6 LEMON THYME

1 **Rosemary** (*Rosmarinus officinalis*). A native of the Mediterranean region and Asia Minor, rosemary succeeds best in the milder and sunnier parts of Britain. It is a sweet-scented cottage-garden favourite, cultivated in the British Isles for more than 400 years and well established in folk-lore. There is an old West Country saying: 'Rosemary will not grow well unless where the mistress is "master"', which is reputed to have caused some husbands to injure the plant secretly in order to destroy the evidence of female domination. Rosemary is of minor importance as a culinary herb, better used fresh, for flavouring meat, poultry, savoury dishes and salads. Its principal commercial value is as the source of an essential oil which is used medicinally, as an ingredient of soap liniment, and in perfumery, shampoos, etc. Most of the oil is produced in Spain, the south of France, and other Mediterranean countries, but the oil distilled in Britain is considered to be superior.

Rosemary is an erect bushy shrub, up to 7 feet high. Its evergreen leaves are dark green above, white hairy beneath, up to 2 inches long and recurved at the margins. The violet-blue or whitish flowers (1A) are borne in small axillary racemes in April and May and sporadically at other seasons. The calyx and corolla are 2-lipped, the latter about $\frac{1}{2}$ inch long, enclosing 2 stamens.

2 **Basil** (*Ocimum basilicum*). Occurring in tropical Asia, Africa, and Pacific Isles, yet hardy enough to be grown in the open in the British Isles, basil has long been used as a culinary herb, although less popular now than it used to be in this country. The aromatic leaves add flavour to soups, ragoûts, and sauces. They have also been used in sausages, salads, and cups. Basil is an erect annual, about 18 inches high, bearing ovate, toothed or entire, long-stalked leaves, 1 to 3 inches long. The flowers are white or purple-tinged, up to $\frac{1}{2}$ inch long, borne in whorls in simple, terminal racemes. A form with deep purple leaves is sometimes grown as an ornamental plant, but its leaves are also aromatic and may equally well be used for culinary purposes.

Bush Basil (*Ocimum minimum*) is a compact plant, 6 to 12 inches high, with leaves smaller than those of the previous species, from which it was probably derived.

3 **Summer Savory** (*Satureja hortensis*). This is a native of the Mediterranean region, well-known to the Romans and noted by Virgil as being one of the most fragrant of herbs and recommended both for culinary use and for planting near bee-hives. The leaves and tender shoots are used, with other herbs, in stuffings for turkey and veal and in meat pies and sausages. The Romans used vinegar flavoured with savory and other herbs as we use mint sauce. Fresh sprigs of savory boiled with peas or new potatoes make a change from the usual mint.

Summer savory is an erect, bushy, rather densely pubescent annual, about 6 to 18 inches high. The linear-oblong, tapering, indistinctly-stalked, opposite leaves are $\frac{1}{2}$ to $1\frac{1}{2}$ inches long. The pale lilac flowers, about $\frac{1}{8}$ inch long, are arranged in whorls which are close together near the apex of the spike, more distant below.

4 **Winter Savory** (*Satureja montana*) is a native of southern Europe and North Africa, with the same culinary uses as Summer savory, from which it differs in several respects, as appears in the following description. It is a hardy perennial sub-shrub, about 12 to 15 inches high, with a small but distinct ridge around the stem between opposite leaf bases. The flowers are larger, with a white or pale purple corolla up to $\frac{2}{8}$ inch long, borne in slender, terminal panicles.

5 **Balm** or **Lemon Balm** (*Melissa officinalis*). Native in the Mediterranean region, and not uncommon as a garden-escape in southern England, balm is grown for its leaves, lemon-scented when crushed and with a lemon-like flavour. Formerly, balm enjoyed a considerable reputation as a medicinal herb, taken as balm tea or as one of the ingredients of Carmelite water, to remedy flatulence, reduce fever, or increase perspiration. Whilst no longer of medicinal importance, balm is still used in herb teas and wine cups, and for making home-made wine.

Balm is a vigorous, perennial herb, 1 to 2 feet high, with rather long-stalked, ovate, toothed or deeply crenate leaves. The white or pinkish flowers are borne in axillary whorls. The toothed calyx and the campanulate corolla are both 2-lipped and the 4 curved stamens are shorter than the corolla.

LIFE SIZE   *FLOWER DETAILS* × 3

1 ROSEMARY   1A Flower detail   2 BASIL   2A Flower detail
3 SUMMER SAVORY   4 WINTER SAVORY
5 LEMON BALM

AROMATIC COMPOSITES

1 Tarragon (*Artemisia dracunculus*). Its leaves are used for seasoning fish sauces, pickles, and salads, and for making tarragon vinegar, which is an ingredient of *sauce tartare* and French mustard. The plant is a native of southern Europe and thrives best in warm, rather dry conditions. It is a bushy perennial herb, up to 4 or 5 feet high if left to grow naturally. The slender, branching shoots, bear smooth, olive green, thin, narrow leaves. The small, whitish-green flower-heads, $\frac{1}{8}$ to $\frac{1}{6}$ inch across, are arranged in racemose panicles. They bear florets of two kinds, female and bisexual, the latter being functionally male. Fertile seed is rarely produced and the plants are usually propagated either by cuttings or by division. The essential oil contained in fresh tarragon disappears when the herb is dried, so the plants are sometimes lifted in autumn and replanted under glass, to provide a continuous supply of fresh green leaves during the winter.

Russian or **False Tarragon** (*Artemisia dracunculoides*) is very similar in appearance but has slightly brighter green, less smooth leaves. It is an inferior substitute for the 'true' or 'French' tarragon.

2 Southernwood (*Artemisia abrotanum*). A herb tea is made by infusing the leaves in boiling water — about 1 oz. of herb to 1 pint. It has an aromatic scent and an agreeable flavour. The herbalists regard it as a stimulant tonic and an antiseptic potent against intestinal worms.

Southernwood or Lad's Love is probably a native of southern Europe. It is a much-branched sub-shrub, 2 to 5 feet high, with finely divided, greyish-green leaves with an apple-like scent. The small, yellow flower-heads, borne in a loose panicle, are rarely produced in this country.

3 Tansy (*Tanacetum vulgare*). This old-fashioned herb has traditional culinary and medicinal uses. The leaves and young shoots can be used for flavouring puddings and omelettes. 'Tansy cakes', made with eggs and young Tansy leaves, used to be eaten at Easter, perhaps as a symbol of the bitter herbs eaten by the Jews at the Passover or perhaps, as one 17th-century author stated, because of their wholesomeness after the salt fish consumed during Lent and because they counter-acted the ill-effects which the 'moist and cold con-stitution of winter has made on people . . . though many understand it not, and some simple people take it for a matter of superstition to do so'. Tansy tea, made by infusing the herb in boiling water, was formerly much used as a tonic, stimulant drink, and for treating children who had 'worms'. The root, preserved in honey or sugar, was recommended as a remedy for gout. The essential oils, from which tansy derives its aroma and flavour, may be beneficial in small quan-tities, but if taken in excess they are violently irritant. Tansy is a common plant of grassland throughout Europe, occurs in Siberia, and has been introduced in North America and New Zealand. Although less often used nowadays, it is still fairly frequently seen in cottage gardens. It is an erect perennial, up to 3 feet high, with pinnate, deeply toothed, dark green leaves, up to about 9 inches long, dotted with glands and very fragrant. The inflorescence is a compound, flat-topped corymb, with numerous, bright yellow, discoid flower-heads, each $\frac{1}{4}$ to $\frac{1}{2}$ inch across. The greenish-white 'seeds' (achenes) are about $\frac{1}{12}$ inch long, with the pappus represented by a short, membranous rim.

4 Alecost (*Chrysanthemum balsamita*). Also known as 'Costmary', 'Balsam Herb', and 'Mace', this herb was formerly used for flavouring ales — hence one of its popular names. The spicy, aromatic leaves have also been used in salads and in pot-pourri. A herb tea, made by infusing them in boiling water, was valued for its astringent and antiseptic properties.

Alecost is a native of Western Asia. In Britain, it often fails to flower and rarely or never sets seed. It is a perennial herb, with entire, bluntly-toothed, oblong or oval leaves, up to 6 inches or more in length, the lower leaves long-stalked, those on the flowering stem sessile and sheathing at the base. The flowering stems are 2 to 4 feet high and bear rather loose clusters of flower-heads, each about $\frac{1}{2}$ inch across and usually yellowish, with a few white florets. The plant is easily propagated, by division of the clumps, in spring or autumn. There are several other useful aromatic members of this family.

Chamomile (*Anthemis nobilis*) yields an essential oil which is light blue when fresh and is used for flavouring liqueurs. Chamomile tea, made by infusing the flower-heads, is an old-fashioned tonic, soothing and good for the digestion. The whole herb has been used for making herb beers. The plant is a native of Europe, local in the British Isles. It is still frequently planted in herb gardens and is sometimes used instead of grass for lawns.

Chamomile (*Matricaria recutica*) is also used for herb teas and for flavouring liqueurs. It is native from southern Europe and western Asia to India and is a locally abundant weed of light arable land and waste places in the British Isles.

LIFE SIZE

1 TARRAGON   2 SOUTHERNWOOD   3 TANSY
4 ALECOST

HERBS: UMBELLIFERS GROWN FOR THEIR LEAVES

1 **Parsley** (*Petroselinum crispum*). Probably native in southern Europe, this familiar fragrant herb is now found escaped from cultivation and more or less naturalized in many temperate regions, including the British Isles. It has been grown in Britain since the 16th century. Early cultivated forms had leaves which were finely divided but not curled and crisped like those of the more popular modern cultivars. Plain-leaved parsley is still grown on the Continent and plants may occur as rogues in some stocks of curly parsley in this country. The leaves are used whole when fresh, for garnishing a variety of dishes and for window-dressing in butcher's and fishmonger's shops. Large quantities are used chopped, fresh or dried, to flavour sauces, soups, salads, omelettes, and stuffing. They are a valuable source of vitamin C.

Parsley is a glabrous biennial. In its first year, it produces a rosette of ternate to pinnate, long-stalked, bright green leaves, 6 to 11 inches long, their segments usually curled and crisped in most varieties. In its second year, it produces solid, erect flowering stems, up to 2 feet high, surmounted by flat-topped, compound umbels of small yellow flowers (1A). The fruits are characteristic of the family Umbelliferae, consisting of pairs of small, one-seeded carpels, joined to a central axis, and falling apart at maturity.

2 **Dill** (*Anethum graveolens*) is usually grown for its fruits as described on page 138, but the young leaves are also used for flavouring soups and sauces, etc., and, like the seeds, in pickling cucumbers and gherkins. The leaves are finely divided, like those of fennel.

3 **Chervil** (*Anthriscus cerefolium*) is native to the Caucasus, Western Asia, and South and Central Russia, but naturalized in many other areas, including the British Isles. It is a slightly hairy annual, 1 to 2 feet high, with 3-pinnate leaves. The small, white flowers, each about $\frac{1}{12}$ inch across, are borne on finely hairy stalks, in small umbels, $\frac{3}{4}$ to 2 inches in diameter. The fruit is comparatively large, about $\frac{1}{2}$ inch long, narrow oblong-ovoid, with a slender, ridged beak (3A).

The cultivated form of Chervil has leaves which are curled, rather like parsley. They are used when fresh, in salads and soups and for garnishing. Chervil is rarely grown commercially but is sometimes seen in private gardens.

4 **Samphire** (*Crithmum maritinum*) is native on seacliffs, and sometimes on sand or shingle, around the western and southern coasts of the British Isles and also on the Black Sea, Mediterranean, and Atlantic coasts of Europe. There is a reference to it in Shakespeare's *King Lear*: 'Half-way down hangs one that gathers samphire, dreadful trade.' For those who wish to try it, without the hazards of climbing more than one cliff, seed collected in autumn should be sown immediately in a light, well-drained soil, preferably in crevices at the foot of a wall where they will get plenty of sunshine. The aromatic fleshy young leaves, salted, boiled, and then covered with spiced vinegar, are an old-fashioned pickle, seldom seen nowadays.

Samphire is a glabrous perennial, 6 to 16 inches high. Its compound leaves have fleshy, tapering cylindrical segments, 1 or 2 inches long. The yellowish-green flowers, about $\frac{1}{2}$ inch across, are borne in terminal compound umbels, each with a whorl of lanceolate bracts. The fruit is ovoid, about $\frac{1}{4}$ inch long, and rather corky.

5 **Sweet Cicely** (*Myrrhis odorata*) occurs in the British Isles, in woods and grassy places, where it may be native but is probably often a well-naturalized escape from cultivation. The whole plant has a sweet, aromatic odour and flavour, reminiscent of anise. In former days it was cultivated for culinary and medicinal purposes and it is still grown in herb gardens. The whole plant may be eaten boiled or the leaves used in salads. The seeds may be used for flavouring purposes. Like several other umbellifers, Sweet Cicely is useful for disorders of the digestive system, but it is no longer used in modern medicine.

Sweet Cicely is an erect perennial, 2 to 3 feet or more in height, with large, 2 – 3 pinnate leaves. The white flowers are of two kinds, short-stalked male and longer-stalked hermaphrodite flowers, in compound umbels. The fruits are large, about $\frac{3}{4}$ inch long, strongly ribbed and glossy brown when ripe.

6 **Lovage** (*Levisticum officinale*) is an old-fashioned vegetable and herb garden plant, formerly more widely used as a vegetable (blanched, like celery) and as a herbal tea. The young stems were candied, like those of angelica. It is a stout perennial, 3 to 4 feet high, with large, bipinnate leaves. The yellowish flowers are borne in compound umbels. It is not a native plant, but is naturalized in some places.

Scotch Lovage (*Ligusticum scoticum*) is a native plant in northern Britain, occasionally used as a vegetable. It can be distinguished by its ternately pinnate leaves, their segments toothed in the upper half and often lobed. Its flowers are greenish-white sometimes flushed with pink.

LIFE SIZE

1 PARSLEY   1A Fruits and flowers   2 DILL   3 CHERVIL   3A Flowers and fruits
4 SAMPHIRE   5 SWEET CICELY   6 LOVAGE

UMBELLIFERS GROWN FOR THEIR LEAF STEMS

1 **Celery** (*Apium graveolens*). Wild celery is a rather local plant in the British Isles in moist places, especially near the sea. It also occurs in central and southern Europe and from western Asia to the East Indies, as well as in Africa and in South America. The cultivated celery, distinguished as var. *dulce*, is occasionally found by roadsides, as a garden escape. Cultivated celery, as we know it today, is a relatively new crop, developed by the French and the Italians, and introduced into the British Isles only in the late 17th century. Seedlings are usually raised under glass in boxes, and planted out in the open when they are a few inches high. The best quality crops, with white leaf-bases, are produced by planting in rows, so that earth can be drawn up on each side of the row every 2 or 3 weeks in order to 'blanch' the plants. Some varieties, e.g. 'Golden Self-Blanching' and 'American Winter Green' (1B), may be grown without blanching. Celery is a particularly useful vegetable, as it can be eaten raw, in salads and *hors d'œuvres*, or as a cooked vegetable, or in soup. The 'seeds' are also used for flavouring purposes.

Celery is a biennial which is used for culinary purposes in its first year, when it forms an upright rosette of leaves with closely appressed, succulent leaf-stalks — the principal part which is eaten. During its second year, the plant produces a tall flowering stem with terminal and axillary umbels of small, greenish-white flowers (1A).

2 **Angelica** (*Angelica archangelica*). A native of northern Europe, Greenland, Iceland, and central Russia, this plant is cultivated and naturalized in many other parts of Europe, including the British Isles. Pieces of the young stems and leaf-stalks, crystallized with sugar, are used by confectioners for flavouring and decorative purposes, their bright green colour giving a festive finishing touch to many cakes and other confections. The main source of supply for crystallized angelica is the south of France. The roots and seeds are rich in essential oils. The former are used, with juniper berries, in making gin, and the seeds in vermouth and chartreuse. The roots, stems, and leaves are distilled individually to produce essential oils.

Angelica is a vigorous biennial, easily grown from fresh 'seed'. It has green stems and large 2- to 3-pinnate leaves with large, oblique, somewhat decurrent, leaf segments. Plants grown for their stems or for their roots are harvested in their first year. If allowed to grow another year, they produce a flowering stem 4 to 6 feet tall, with umbels of greenish-white or green flowers. The Garden Angelica should not be confused with our native Wild Angelica (*Angelica sylvestris*), which usually has purplish stems and white or pink flowers.

3 **Florence Fennel** (*Foeniculum vulgare* var. *dulce*). The English name indicates the country of origin of this vegetable, which, in Italy, is known as '*finnochio dolce*'. It is a short stocky plant, about 1 foot high, with greatly swollen leaf-bases (3), forming a kind of false bulb, often as big as a fist, which is the part eaten. Seed is usually sown in rows about 18 inches apart and thinned to a spacing of 6 inches. As the leaf-bases enlarge, they are slightly earthed-up. Florence fennel has a mild, sweet flavour, somewhat resembling celery. It may be eaten raw, sliced thinly in salads, or cooked in water or stock and served with an oil and lemon juice dressing or a butter sauce. Apart from its relatively small size and swollen base, Florence fennel bears a general resemblance to ordinary fennel, in its much divided leaves with thread-like segments arranged in various planes. The flowering stems are about 2 feet high, bearing umbels of yellow flowers.

1 CELERY blanched leafstalks   1A Flowers   1B 'AMERICAN WINTER GREEN' plant
2 ANGELICA stem   2A Plant
3 FLORENCE FENNEL leaf bases   3A Plant

149

COMPOSITE SALAD PLANTS: LETTUCE, ENDIVE, CHICORY

1-3 Lettuce (*Lactuca sativa*). The garden lettuces have been cultivated for so long that their origin is uncertain, although the Near East, the Mediterranean region, or Siberia have been suggested as possible sources. Lettuces are composed of over 95 per cent water, but, like other green vegetables, they contribute vitamin A to our diet. They are the best known of all the green salad plants but they can also be used cooked. Bolted lettuces, eaten as a vegetable in some quantity, have been known to cause coma.

Lettuces are commonly classified as either 'cabbage' or 'cos', and according to their growing season as 'winter' or 'summer'. Summer lettuces are grown in the open, preferably in sandy or loamy soil, and thrive best in rather cool situations. The modern winter lettuces need the protection of a greenhouse or a frame, at least in cool climates. The older winter lettuces which could be grown out-of-doors are rarer nowadays, as they cannot compete with the tenderness of crops grown under glass.

Cabbage Lettuces (1 – 2) have roundish or somewhat flattened heads; they vary in texture from the soft 'All the Year Round' to the crisp 'Webb's Wonderful' (1). 'Buttercrunch' (2) is a new fairly crisp variety which is reputed to last longer before running to seed.

Cos Lettuces (3) have longer leaves, forming relatively elongated, upright heads. The older cos lettuces had to have their heads ringed with raffia to produce good quality hearts. Modern varieties, such as 'St. Albans' (3) produce reasonably compact hearts without this attention, although for show purposes they are still usually loosely tied.

The lettuce is an annual or biennial belonging to the Daisy family (Compositae). If allowed to run to seed (1A), it produces an erect, branched flowering stem, 1 to 3 feet high, bearing small, pale yellow flower-heads. Its light or dark greyish-brown 'seeds', about $\frac{1}{8}$ inch long, bear a 'pappus' of soft, white hairs, which acts like a parachute, aiding wind-dispersal.

4 Endive (*Cichorium endivia*). This salad plant, more popular on the Continent than in Britain, is probably a native of southern Asia and northern China. Endive is sown in a rich, well-drained soil, in April and May for autumn use and from June to early August for winter and early spring crops. The latter are usually sown out of doors and transplanted into frames in autumn when frost threatens. In mild localities, such as Cornwall, this precaution is unnecessary. As green endive leaves are very bitter, the leaves are blanched to a pale yellow colour to make them palatable, by covering them, usually with litter, to exclude the light. In early autumn, the blanching process takes 5 to 10 days, but in winter up to 20 days.

Endive is an annual or biennial plant which produces a dense rosette of glabrous leaves. The most popular varieties are those with leaves which are much divided and curled, for example, 'Ruffec Green Curled'; but 'Round-leaved Batavian', with leaves toothed and wavy rather then curled, has a particularly fine flavour. Endive, if left to flower, bears a branching, leafy, flowering stem, up to 3 feet high, with pale blue flower-heads about $1\frac{1}{2}$ inches across. The 'seed' is three to six times as long as its pappus.

5-6 Chicory (*Cichorium intybus*). Native from Europe to western Asia and central Russia, chicory is also locally common, but probably often as an escape from cultivation, in England and Wales, especially on chalky soils. It has been used as a salad since ancient times. Some varieties are grown for their large roots, which, when dried, roasted, and ground, are used blended with coffee (*see* p. 110). The salad chicories are, like endive, usually blanched to make them less bitter, and the leaves, whole or shredded, often dressed with oil and vinegar. The blanched hearts are also served as a cooked vegetable.

The seed is sown in the open in May or June, preferably in a light, fertile soil, and the seedlings thinned to about 9 inches apart. For forcing and blanching, the plants are lifted, from late September to mid-November, and replanted in special forcing beds, in heated sheds, or in greenhouses. The leaves are cut off 1 inch above the root and the crowns are covered with dry, light soil, peat, or sand. Young shoots grow within the covering material, which shields them from light and therefore prevents the development of the green chlorophyll pigments. The variety usually grown commercially, for blanching, is 'Witloof', which has broad leaves and wide midribs. When blanched, it forms a compact head, similar to a cos lettuce in shape. 'Sugar Loaf' (6) is a rather similar modern variety. 'Rossa di Verona' (5) has leaves bearing irregular brownish-red markings which become brighter red on blanched heads.

Chicory is a perennial herb. It produces a branching flowering stem, about 3 feet high, bearing bright blue flower-heads (6A). The 'seed' is crowned by a small pappus, only about $\frac{1}{10}$ of its length.

CRUCIFER SALAD PLANTS

1-2 Watercress (*Nasturtium officinale and N. microphyllum × officinale*). Usually eaten raw as a salad, watercress has a pleasantly pungent flavour, rather like mustard and cress. It is a valuable source of vitamins and also has a fairly high protein content. In his *Herball*, dated 1597, John Gerarde wrote: 'Water Cresse being boiled in wine or milke, and drunk for certaine daies togither, is verie good against the scurvie or scorbute.' He also recommended it to be eaten boiled 'in the broth of flesh' to cure young maidens of the greensickness.

Watercress was formerly gathered from the wild and it was not until the early 19th century that serious attempts were made to cultivate it as a commercial crop. It requires a carefully selected site and a supply of clear, uncontaminated water. Given good conditions, a bed may yield as many as ten crops a year and could go on yielding for ten years, although frequent replanting is more usual. There are no named cultivated varieties, but selected cultivated strains are often larger-leaved and more freely branching than the average wild watercress. There are two main types grown commercially in the British Isles.

Green Watercress (*Nasturtium officinale*) remains green in the autumn and is susceptible to frost damage in winter and spring. Its fruit (a siliqua) bears two distinct rows of seeds in each cell.

Brown or Winter Watercress is a hybrid between the former and another wild species, One-rowed Watercress (*Nasturtium microphyllum*). It turns purplish-brown in autumn and is less affected by frost. Its fruits are usually deformed and contain few or no good seeds.

One-rowed Watercress also turns purplish-brown in winter. Its seeds are arranged approximately in one row in each cell.

3 Mustard (*Sinapis alba*). Grown mainly for eating raw, in salads and sandwiches, the plants are harvested when they are a few inches high, when only the first pair of seed-leaves (the cotyledons) have expanded. The crop is usually grown in greenhouses, which enables it to be produced all the year round if a temperature of 50° – 60°F. can be maintained. The seed is sown on the surface of the soil, on firm, level beds. It is watered with a fine spray, then covered with steam-sterilized wet sackcloth, which is sprayed to keep it moist and is removed when the seedlings are 1 – 1½ inches high — in about 4 days in spring and autumn, 6 to 7 days in winter. The yellowish seed-leaves turn green within 2 or 3 days and the crop is then cut. It is usually marketed in small boxes or punnets, often packed together with cress.

For home use, small quantities of mustard and cress are often grown on wet flannel on a dish, covered to exclude the light and to keep the seedlings moist.

Much of the so-called 'Mustard' grown commercially is not the true White Mustard (*Sinapis alba*), but Rape (*Brassica napus*). Its seedling leaves are a rather more intense shade of green than those of White Mustard; it keeps better in warm weather and its seed is usually cheaper. If allowed to go to seed, the difference between the two plants is apparent in several characteristics, especially in the fruits, which are bristly hairy, with a flattened sabre-like beak in *Sinapis alba*, non-hairy with a slender tapering beak in *Brassica napus*. The strains of Rape used for salads are selected for the purpose. Other strains of the same species are a source of edible oil (*see* p. 25).

4 Cress (*Lepidium sativum*) is eaten, like mustard (and often with it) in the seedling stage. Cultural conditions for the two plants are similar, except that cress takes 3 or 4 days longer to germinate and allowance has to be made for this if the two crops are to be cut at the same time. Mustard and cress is sometimes sown in the plastic or waxed punnets or trays in which they are to be sold, usually in peat which has been treated with a nutrient solution.

Cress is believed to be a native of western Asia but is cultivated and naturalized in many parts of Europe, including the British Isles. It is quite distinct from both mustard and salad rape, in appearance. Its cotyledons are deeply 3-lobed and its mature leaves are pinnate or bipinnate. Some varieties of cress have crisped leaves, like those of parsley. They are sometimes used for garnishing. Cress is an annual, with a single or sparsely-branched, erect stem, 8 to 16 inches high. Its flowers have 4 white or reddish petals. Its fruit is a silicula (4A), a flattened, elliptical pouch, notched at the apex and up to ¼ inch long, usually containing one seed in each cell.

Winter Cress or **Land Cress** (*Barbarea verna*), is a useful but rarely-grown salad plant. It can be picked throughout much of the winter, if sown in late summer and given the protection of a cloche, frame, or dry leaves during severe weather. It is biennial, normally grown as an annual, native of Europe. Its flowering stem is 2 to 3 feet high, with pinnately lobed leaves, bright yellow flowers, and a narrow, cylindrical, short-pointed siliqua, 1½ – 2½ inches long.

5 Rocket (*Eruca sativa*) is another rarely-grown salad plant. The seed is the source of an oil which is used as a substitute for rape oil (*see* p. 25). The plant is an annual, native of the Mediterranean region and naturalized or casual in the British Isles and elsewhere. Its erect stem, up to 2 feet high, bears pinnately-lobed leaves, with a large terminal lobe (like those of radishes or turnips), the upper leaves less deeply lobed or toothed. The terminal inflorescence bears white or yellowish flowers, with deep violet or reddish veins. The fruit is rather like that of White Mustard, but lacking bristly hairs — a cylindrical siliqua, with a broad, flattened, sabre-like beak.

LIFE SIZE

1 CULTIVATED WATERCRESS   2 WILD WATERCRESS

3 MUSTARD seedlings   3A Flowers   4 CRESS seedlings   4A Flowers and fruits

5 ROCKET

ORIENTAL BRASSICAS AND 'CHOP SUEY GREENS'

The regions of eastern Asia have their own distinctive culinary vegetables, some of which can be grown successfully in the British Isles and other temperate countries. Seeds of several Chinese and Japanese vegetables are stocked by some British seed merchants.

1 **Pak-Choi** (*Brassica chinensis*) is also known as 'Chinese Cabbage', although it is more closely related to rape and swede than to the European cabbages. The leaves can be eaten raw, in salads, or cooked, as 'greens'. The plant does not form a heart and in appearance it resembles chard or spinach beet rather than a cabbage. The broad, smooth-edged leaves, tapering gradually into a narrow-margined stalk, may be 10 to 20 inches long when well grown. Pak-choi does best when sown in July or August, to produce an autumn crop. If sown earlier in the year, it runs to seed too rapidly. The yellow flowers are borne in a terminal raceme. Some of the older flowers usually overtop the unopened buds. The long-stalked, slender, beaked fruit is $1\frac{1}{4}$ to $2\frac{1}{2}$ inches long.

2 **Pe-Tsai** (*Brassica pekinensis*) is grown mostly for use as an autumn and winter vegetable, equivalent to 'spring greens'. Pe-tsai has soft green, prominently veined leaves, mostly obovate, with wavy, variously-toothed margins. The midrib is light green. The blade tapers about the middle and is continuous to the base as a broad, jagged wing on each side of the petiole. The rather loose, elongated head bears some resemblance to a cos lettuce. In China, solid heads of different shapes are produced. Pe-tsai is an annual and if sown too early in the year it tends to run to seed. The light yellow flowers, about $\frac{3}{8}$ inch long, are crowded in a terminal raceme. Some of the older flowers overtop the unopened buds. The fruit is $1\frac{1}{4}$ to $2\frac{1}{2}$ inches long, rather stout, with a short beak.

3-4 **Wong Bok** and **Chihli.** These are two more varieties of *Brassica pekinensis*, differing from *Pe-tsai*, and from each other, in minor distinctions of colour, degree of hairiness, and the varying sharpness of the teeth around the edges of their leaves. Their uses are the same.

5 **Shungiku** (*Chrysanthemum coronarium*). This is one of the many chrysanthemums grown for ornamental purposes in the western world where its common name is 'Garland Chrysanthemum'. In China and Japan, the young plants are used as a cooked vegetable, generally in combination with other foods. To the western palate, their flavour may seem rather strong. The Garland Chrysanthemums grown for ornament are 2 to 3 feet high, with leaves which are divided into deep, pointed lobes. The kind grown for use as food has fleshier leaves, with a large terminal lobe and much smaller lobes near the base. It is sometimes considered to be a distinct species (*Chrysanthemum spatiosum*). Both kinds have yellow or yellowish-white flower-heads.

QUARTER LIFE SIZE

1 PAK-CHOI

2 PE-TSAI   3 WONG BOK   4 CHIHLI
5 SHUNGIKU

155

THE EUROPEAN BRASSICAS
(VARIETIES OF BRASSICA OLERACEA) (1)

It is difficult to conceive that the wild cabbage which grows on our coastal cliffs has given rise to so many distinctive cultivated races, such as cabbages and savoys, Brussels sprouts, kales, broccoli, cauliflowers and kohlrabi. Though so different in their vegetative growth, yet when they 'run to seed', their common ancestry is apparent in the similarity of their inflorescences, flowers, and fruits.

1 **Wild Cabbage** (*Brassica oleracea*). (Some botanists regard it as a distinct species, *Brassica sylvestris*.) The wild cabbage is apparently native in southern England and Wales and on the coasts of the Mediterranean and the Adriatic, from north-east Spain to northern Italy. In other areas, it is believed to be an escape from cultivation. Several cultivated varieties — cabbage, cauliflower and broccoli — were known to the Ancient Greeks and the Romans. The Saxons and the Celts grew cabbages in northern Europe.

It is a glabrous biennial or perennial plant, with a rather woody, more or less decumbent stem. Its glaucous, blue-green leaves are few, compared with most of its cultivated relatives. The lower leaves are stalked and fairly large, with irregularly sinuate margins and often with a few small lobes near the base. The 4 pale yellow petals each has a broad, horizontal limb, tapering abruptly into the narrow, vertical claw. The stamens are characteristic of the Wallflower family (Cruciferae): 2 outer short stamens and 4 inner long stamens. The fruit is a siliqua, a narrow capsule, 2 to 4 inches long, with a short, usually seedless, beak, opening from below by 2 valves to expose the seeds which are borne on a central wall.

2 **Kales** are hardy 'winter greens', often grown in home gardens as a standby for use in hard winters when other brassica crops may be damaged. Compared with other 'greens', some of the kales are rather strong flavoured and less succulent when cooked, but different cultivars vary greatly in this respect. Kales usually have a simple, erect stem, bearing large leaves which may resemble those of wild cabbage, although such forms are used mainly for feeding livestock. The Curly Kales, which have their leaves curled and crimped, like those of parsley, are more popular for human consumption. In recent years, there has been increasing interest in kales with purple or silver-variegated leaves, for floral arrangements and garnishing as well as for culinary use. Some kales branch freely, in spring, producing tender young shoots more delicate in flavour than the old leaves.

3 **Cabbages** are characterised by their greatly enlarged terminal bud, which forms the familiar cabbage. Often scorned by gourmets, but perhaps because they are frequently cooked too long and in too much water, the cabbages are one of our oldest green vegetables. From the growers' point of view, cabbages may be divided into two categories: those suitable for spring sowing, which may bolt if sown in autumn; and those sown in autumn, which may fail to 'heart' if sown in spring. In each category, there are round-headed and conical forms. There is a further division into Spring Cabbage, which may be cut either when semi-hearted, as 'spring greens', or when hearted; Summer to Autumn cabbage, cut from June to October; and Winter Cabbage.

4 **Red Cabbage** differs from other cabbages most obviously in colour, mainly due to the content of anthocyanin pigments. Red cabbages are used chiefly for pickling.

5 **Spring Cabbage.** This term is used for young cabbage, regardless of season. Increasing popularity in recent years has made it worthwhile for growers to extend the period of production; by commencing sowing in June, they have made 'Spring Cabbage' available from October to June.

6 **Savoys** are cabbages with wrinkled leaves, generally hardy and frost resistant, in season from September to March. They tend to be milder in flavour than the smooth-leaved winter cabbages and make excellent shredded cabbage salad or 'Cole slaw' — a term which is presumably derived from the old fashioned 'Cole-worts', small cabbages formerly much grown for use as 'greens'.

7 **Brussels Sprouts** are grown for their dense, compact, axillary buds (like miniature cabbages), which are borne close together all along a tall, single stem and have a distinctive and popular flavour. They are suitable for quick-freezing, and large quantities are processed in this way. The terminal bud and rosette of leaves is not compact and is of little commercial importance, although 'sprout tops' have a limited sale. In recent years, F^1 (first cross) hybrid sprouts have been recommended for their uniformity and heavy yield.

8 **'Flower Cabbages'.** These are ornamental cabbages, recently introduced from Japan, attractive in garden borders and popular for floral arrangements. The head is loose, with the inner leaves bright pink or pale yellow and the outer leaves beautifully variegated.

PLANTS × ⅛   *DETAILS* × ½

1 WILD CABBAGE leaf and flower details   2 KALE   2A CURLY KALE leaf detail
3 ROUND CABBAGE   4 RED CABBAGE   5 SPRING CABBAGE   5A Leaf detail
6 SAVOY   6A Leaf detail   7 BRUSSELS SPROUTS   7A Detail of sprout
8 FLOWER CABBAGES

157

THE EUROPEAN BRASSICAS
(VARIETIES OF BRASSICA OLERACEA) (2)

1 **Cauliflower.** This familiar vegetable is sometimes served raw when young (divided and immersed in a French dressing), but is more often boiled and served with a bechamel sauce, or *au gratin*, or in various other ways. Cauliflower is also a component of various pickles. Horticulturally, those varieties which mature in summer and autumn are termed 'cauliflowers', whereas the slow-maturing winter cauliflowers are often known as 'broccoli', but this distinction is lost in the markets, where both groups are usually sold as cauliflowers. In Britain, summer and autumn cauliflowers are grown all over the country, but winter cauliflowers are grown mainly in mild coastal areas, such as Cornwall, where frost damage is unlikely. In winter, large quantities of cauliflowers are imported from Italy and Brittany.

Cauliflowers normally produce a single stem, bearing a large, swollen, roundish flower-head, consisting of a tightly-packed mass of undeveled, white or creamy-white flower buds, known as the 'curd' (1A). This flower-head nestles in the surrounding leaves, which are erect or spreading in summer cauliflowers. In winter cauliflowers, the innermost leaves are wrapped over the curd and help to protect it from weather damage.

2 **Sprouting Broccoli** is similar to the cauliflowers in the structure of its flower-heads, but instead of producing a single head it produces a rather loose terminal cluster of flower-heads (2A) on one or several branches, and a large number of smaller heads in the axils of the leaves lower down the branches. Purple and white varieties are available, the purple (which turns green when boiled) being the more popular. The crop is sown in April and is ready for cutting in the following March or April, early varieties maturing a few weeks earlier than late varieties.

3 **Green Sprouting Broccoli** or **Calabrese** is an Italian vegetable, grown commercially mainly for quick-freezing or canning. Being frost-tender in Britain, it is usually sown under glass in April and not planted out until June. It matures rapidly and is ready for picking during August and September. The plant forms a rather loose and fairly large terminal head bearing clusters of green flower buds. The flowering stems are often fasciated (i.e. two or more parallel stems grow together, like the twin barrels of a shot-gun). When the terminal head is cut, side branches continue to develop, producing a succession of small heads. Calabrese is usually eaten boiled, like broccoli, but also as a purée and in soups.

4 **Kohlrabi** is sometimes called 'Turnip-rooted cabbage', which is a good, brief description of a plant which looks something like a turnip but is botanically more closely related to the cabbage. The turnip-like globe of kohlrabi is actually the swollen base of the stem, not the root. It bears leaves, just like a normal stem. Kohlrabi can grow as big as a large orange, but it is best eaten before it is fully grown. It is used as a boiled vegetable and as a food for livestock in some European countries, but it has never become very popular in the British Isles. Seed sown in early spring, the seedlings thinned out to about 9 inches apart, will yield a summer crop, and successional sowings are made up to the end of July for autumn and winter supplies. There are two kinds, green and purple (4A).

PLANTS × ⅛   DETAILS × ½
1 CAULIFLOWER   1A Detail of inflorescence
2 PURPLE SPROUTING BROCCOLI   2A Detail of flowering shoot
3 CALABRESE detail   4 KOHLRABI   4A Detail of stems

OTHER LEAF VEGETABLES

1 **Spinach** (*Spinacea oleracea*) is widely cultivated in the temperate regions of the world. Its leaves are eaten as a boiled vegetable (preferably cooked with little or no water) and in soups. Large quantities are canned and frozen. Spinach is much richer in protein than other leaf vegetables, and it also has a high vitamin A content.

Spinach is an annual, belonging to the Goosefoot family (Chenopodiaceae). There are two main groups of varieties: the round-seeded or summer spinach (illustrated), and the prickly-seeded or winter spinach. Round-seeded varieties are sown in spring and summer, for picking as soon as the leaves are ready, before the plants run to seed. They are also sown in late summer, for picking from October to May. Prickly-seeded spinach, as the name suggests, has spiny 'seeds' (botanically, fruits). It forms a more spreading, branching plant, bearing broadly triangular leaves. It was formerly thought to be hardier than the round-seeded varieties but this has been disproved and the latter are now preferred even for winter use.

Spinach plants quickly run to seed, in summer, producing a leafy stem about 2 feet high, with small, green flowers, the male flowers (1A) in clusters on long, terminal spikes, the females often axillary.

2 **Spinach Beet** (*Beta vulgaris*) is closely related to the garden beetroots and sugar beets, but it is grown solely for its leaves, which are used as a green vegetable, like spinach, which it resembles in flavour. The whole leaf can be eaten, including the long, green stalk. Apart from the succulence of its leaves and its relatively unswollen tap-root, spinach beet resembles the other beets (*see* pp. 15 and 171) in its botanical characteristics. It may be sown in spring, for picking in summer and autumn, or in late summer, for picking in late winter and early spring.

3 **Seakale Beet** or **Chard** (*Beta vulgaris*) is very closely allied to spinach beet and is used in the same way. It differs mainly in having a broad, white leaf-stalk, up to 2 or 3 inches across, which is often eaten as a separate vegetable, cooked and served in the same way as seakale, whilst the green blade is used like spinach. To those who find spinach too acid, the milder flavour of spinach beet and seakale beet may be more acceptable.

4 **Orache** (*Atriplex hortensis*) is occasionally grown as a substitute for spinach or sorrel. It is another member of the Goosefoot family (Chenopodiaceae), from temperate Europe and Asia. There are red-leaved (4) and yellow-leaved forms, which are grown as ornamental plants but which can be eaten just like the normal green-leaved form. Orache is a tall annual herb, up to 7 feet high, with long-stalked, triangular leaves, up to 5 inches (rarely 8 inches) in length. Male and female flowers are borne on the same plant. Most of the female flowers have two large bracts but no perianth; about a quarter of them have a perianth but no bracts.

5 **New Zealand Spinach** (*Tetragonia expansa*) is eaten like spinach; but it belongs to the Ice-plant family (Aizoaceae). It is less hardy than the spinach and the spinach substitutes so far mentioned and in the temperate zone it does best in hot summers; in tropical countries it is valued for its drought-resistant qualities. It is a vigorous annual, with spreading stems 2 to 3 feet long, bearing fleshy leaves 2 to 5 inches long. The inconspicuous, yellowish-green flowers are succeeded by a hard, top-shaped fruit, about $\frac{1}{3}$ inch long, crowned with horns.

6 **Amaranthus Spinach.** Several species of *Amaranthus* are locally used as spinach in different parts of the tropics. They are all plants which grow wild; sometimes they appear as weeds in cultivated ground, and are allowed to grow so that their leaves may be picked and cooked, and occasionally a few plants are deliberately sown. Some of them, like the one shown in the illustration which is used in Zambia and neighbouring countries, have red or reddish leaves. There are also certain species of *Amaranthus*, of which *A. leucocarpus* is the most important, grown for their seeds which are used as grain in various parts of the tropics, especially in Central America and the Andean region of South America.

PLANTS × ¼   FLOWERS × 1   DETAILS × 3

1 SPINACH   1A Male flowers   1B Male and female flower details
2 SPINACH BEET   2A Flowers   2B Detail   3 SEAKALE BEET (CHARD)
4 RED ORACHE   4A Flowers   5 NEW ZEALAND SPINACH
6 AMARANTHUS SPINACH   6A Flowers

SOME PLANTS GROWN FOR THEIR YOUNG STEMS
AND LEAF STALKS

1 **Rhubarb** (*Rheum rhaponticum*). Garden rhubarb is believed to have been derived from a wild Siberian species, but it may well be of hybrid origin. The long, fleshy leaf-stalks are usually red (green stalks are less popular). They are used like 'fruit', stewed with sugar, in pies and preserves, and for home-made wine. Forced rhubarb, grown early in the season, is generally preferred as it is tender and less acid than the later, unforced stalks. For forcing, the rootstocks are lifted and left on top of the soil for about two weeks, exposed to the winter weather. They are then placed in the dark, in sheds or beneath greenhouse staging, or covered with barrels. The young leaf stalks develop rapidly and are usually an attractive pale red, with yellowish instead of green leaf-blades. Eating rhubarb leaves as a vegetable has sometimes caused deaths, probably because of their high content of oxalic acid.

2 **Seakale** (*Crambe maritima*) is a native of sea-cliffs, sands, and shingle around the coasts of Western Europe. It was formerly extensively cultivated in Britain but today it is a relatively uncommon vegetable. The blanched leaf-stalks are boiled, like asparagus. They have a nutty, slightly bitter flavour. The plants are grown usually from root-cuttings, but sometimes from seed. They are blanched in winter and early spring either in the open or in hot-houses; a pot or box is placed over the crown and is covered with fermenting manure to create warmth for forcing early crops. Wild Seakale is sometimes blanched where it grows, by covering the crowns with about 18 inches of shingle. The young shoots are cut when their cluster of blanched leaf-stalks is about 5 inches high.
Seakale belongs to the Wallflower family (Cruciferae). The plant has broad, lobed and toothed, bluish-green leaves up to 1 foot long, and flowering stems up to 2 feet high, with white, 4-petalled flowers. The roundish fruit contains a single seed.

3 **Asparagus** (*Asparagus officinalis*) is a native of Europe. In Britain, it has become naturalized in waste places and sand dunes. (A distinct subspecies, ssp. *prostratus*, is a rare native plant occurring on grassy sea-cliffs.) Asparagus has been cultivated since the time of the ancient Greeks, and is usually considered to be a luxury vegetable.
The part eaten is the young shoot or 'spear', which arises from the rootstock between late April and early July and is cut when 9 to 12 inches high. Asparagus is usually poached in water and served simply with butter or maybe a sauce. Part of the crop is quick-frozen and considerable quantities of canned asparagus are imported. In Britain, the chief producing areas are in Suffolk, Worcestershire, Norfolk, Essex, and Lancashire.
Asparagus, a member of the Lily family (Liliaceae), is grown from seed, but the home gardener usually plants two- or three-year-old crowns bought from nurserymen. There are several named varieties, e.g. 'Connover's Colossal', 'Argenteuil', 'Sutton's Perfection', and 'K.B.F.', but the differences between these strains are not clearly defined. Cropping starts when the plants are 3 years old, the later-developing shoots being left to maintain the vigour of the crop in succeeding years. Asparagus shoots are green-tipped or purple-tipped, and white towards the base where the soil has kept the light from them. They are cylindrical, with scale-like leaves, increasingly numerous towards the apex where they are close together, giving a somewhat cone-like appearance. The shoots which are left on the rootstock elongate to form much-branched stems about 5 feet high, bearing clusters of needle-like 'cladodes' (modified branches which function as leaves) in the axils of the scale-leaves. The small, yellowish or pale green male and female flowers are normally borne on separate plants, singly or in groups of 2 or 3, at the junctions of the branchlets. Occasionally, hermaphrodite flowers are produced, in which both pistil and stamens are functional. The fruit is a small, round berry, red when ripe.

4 **Bamboo Shoots,** as an article of food, are the thick pointed shoots which emerge from the ground under a bamboo plant and which would, if left, develop into a new stem or culm (4A). At the emergent stage they are, in many species, soft enough to be eaten after boiling. Bamboo shoots are used as an article of diet mainly in eastern Asia — especially in China, Korea, and Japan where they move into markets in considerable quantity. At the stage at which they are usually cut they may be from 6 inches to 1 foot long and up to 3 inches or more in diameter. The shoots of many of the numerous species of bamboos grown in Asia are eaten, though some are more palatable than others. Some shoots come from wild bamboos. Some come from bamboos which have been planted for quite other purposes, such as providing stems for building — for a certain number of shoots can be removed without depressing the growth of the clump, as all do not usually develop. Sometimes bamboos are deliberately planted to provide edible shoots; *Bambusa vulgaris* is widely used for this purpose, as is *Phyllostachys pubescens*, a Chinese species which is also popular in Japan. As well as being eaten when freshly boiled, the shoots are sometimes salted or pickled.

SHOOTS AND LEAF-STALKS × ½   *PLANTS* × $\frac{1}{12}$

1 RHUBARB leaf-stalks   1A Plant   2 SEAKALE blanched leaf-stalks   2A Plant
3 ASPARAGUS shoots   3A Plants   4 BAMBOO shoot   4A Culms

CARDOON AND GLOBE ARTICHOKE

1 Globe Artichoke (*Cynara scolymus*). The young flower-heads or 'chokes' (1A) have numerous large scales or bracts with fleshy bases (1B). This fleshy base is the part usually eaten. The flower-head may be baked, fried, boiled, stuffed, or served with various sauces or dressings. In Italy, baby artichokes are eaten raw as an appetiser, preserved in olive oil; or they are dipped in batter and fried. The fleshy receptacle (1C) at the base of the young flower-head is also eaten, and the tender central leaf-stalks, blanched, can be used like Cardoons.

The globe artichoke probably originated somewhere in the Mediterranean region, where it was known to the ancient Greeks and Romans. It can be grown from seed, but the seedlings are variable. The named varieties can be grown only from suckers, which are cut off from the base of the parent plant in March and planted in their permanent positions, 2 feet apart, with 4 feet between the rows or groups of plants. They need shading at first, until established, and do best in a deep, rich, but not heavy, loam. Plants grown from suckers often start to produce heads in their first year and more in succeeding years. They are usually replaced by younger plants after 4 years. If cut down early in July, the plants will produce suckers which can be covered and blanched, like Cardoons. The crowns need protecting with straw or bracken during the winter, except in mild districts.

The globe artichoke is a thistle-like plant belonging to the daisy family (Compositae). It is an herbaceous perennial, 3 to 5 feet high, with arching, greyish-green leaves often more than 2 feet long. The leaves are not truly pinnate, but have deep, pointed lobes (1). The globose flower-heads (1A), often 3 inches or more in diameter, consist of a common receptacle (1C), bearing broad, fleshy, green scales (the part eaten), surrounding a central cluster of violet-blue florets. Although usually grown for use as a vegetable, the globe artichoke is a handsome plant worthy of a place in the flower garden. In Britain, varieties are usually listed as 'green' or 'purple', but in France where this vegetable is more popular, several named varieties are grown.

2 Cardoon (*Cynara cardunculus*). This vegetable is grown primarily for its leaf-stalks which are blanched and eaten like celery, in salads, soups and stews. The roots are less often used as a cooked vegetable. The plant is closely related to the globe artichoke, but is not held in such high esteem by gourmets. It, too, is a native of the Mediterranean region.

Cardoons are usually raised from seed, sown in late April in a trench about 1 foot deep and covered with 1 inch of soil. The seeds are sown in groups 20 inches apart and the seedlings are thinned out, leaving one to each station. The plants are blanched when they are full grown, in September; the gardener draws the leaves together, ties them with raffia, wraps brown paper around each plant, and then surrounds it with a hayband about 3 inches thick, covering the whole length of the plant. The plants are banked up with soil, like celery, and should be blanched in about a month. For further protection from frost the ridges are covered with litter; or the plants may be dug up and stored, still in their protective coatings, in a frost-proof place. Considering the amount of labour involved, it is not surprising that cardoons are not a commercial crop and are rarely grown even in private gardens.

The cardoon closely resembles the globe artichoke plant in habit and appearance. The cardoon's leaves are more spiny than those of the artichoke and its flower-heads have spine-tipped bracts.

3 Okra (*Hibiscus esculentus*) is also known in different countries as 'Okro', 'Lady's Fingers', and 'Gumbo'. The plant is an annual, 4 – 6 feet high, with pretty yellow flowers (3A). As it belongs to the Malvaceae, the same family as cotton, it harbours many of the pests and diseases which attack cotton, so that in some cotton-growing countries 'close seasons' when no cotton plants are allowed to be in the ground are also applied to okra. The part which is eaten is the whole fruit or 'pod' (3B). As the pods when ripe are too fibrous to be digested, they must be picked in the immature stage usually about 2½ months after planting. They may be up to 9 inches long. They are eaten either fresh or cooked, are sometimes canned, and can also be dried to conserve them for later use. As they are very mucilaginous, they are often used in tropical cookery to thicken soups and stews; this property and their pleasant taste provide their chief value as a foodstuff, since the actual nutrient content of the pods is not outstanding in any way. The okra is native to tropical Africa. Although not suited to altitudes above about 3,000 feet it is now very widely grown in the tropics and subtropics and is popular as far from the equator as Spain. There are a number of named varieties in different regions.

PLANTS × $\frac{1}{12}$ *DETAILS* × $\frac{1}{2}$

1 GLOBE ARTICHOKE

1A Immature flower-head   1B Cooked flower-head   1C Receptacle

2 CARDOON

3 OKRA   3A Flower   3B Fruits

Several species of *Allium* have been used as food since very early times. Onions and garlic were eaten in Egypt about 3000 B.C., and leeks were eaten by the Israelites before the time of their exodus from Egypt. Today, these and other relatives of the Onion are used, either as separate vegetables or as flavourings for other foods, in most parts of the world.

The Onion belongs to the Alliaceae family, which is intermediate between the Liliaceae and the Amaryllidaceae, in both of which it has been classified in the past. It is a biennial plant, storing food in the bulb during its first year and flowering in its second year. Its leaves and flowering stem are both hollow and the latter is swollen near the middle, tapering towards both ends. The large, globose flower-head bears many greenish-white flowers.

1-4 Onion (*Allium cepa*). Of uncertain origin, onions are widely grown and are exported in large quantities by Egypt, Holland, Spain, the U.S.A., Italy, and other countries. Britain is the leading importer of onions, importing about 200,000 tons annually, despite home production of about 48,000 tons. The culinary uses of onions are extraordinarily numerous. They are eaten raw, fried, boiled, and roasted; in soups, sauces, stews, curries, and a great variety of other savoury dishes; and they are a main ingredient of many pickles and chutneys.

The section of an onion (1) shows that it is a bulb, composed of fleshy, enlarged, leaf-bases. In shape, onions of different varieties may be flattened globose, globose, or oval. In colour, the bulbs vary from an almost silvery-white — like 'White Lisbon' (4) — to pale or dark brown or occasionally red (3). Onions which are to be allowed to grow to their full size may be grown from seed or from 'sets'. The latter are small onion bulbs, specially grown for the purpose. They are produced during the summer and stored, preferably at a temperature of 75°F., during the winter, to be replanted in spring. By using selected varieties and keeping the small bulbs warm during the winter, large bulbs can be produced with a minimum of 'bolting' (starting to flower).

'Giant Zittau', 'Bedfordshire Champion' and 'Up-to-date' all store well. 'Rijnsburger' is high yielding and stores moderately well. 'Ailsa Craig' is popular for sowing under glass in autumn and early spring, and transplanting into the open ground in spring; this ensures a longer growing season and larger bulbs than are obtained from seed sown in the open.

Spanish onions (2) of which there are several varieties, are characteristically large and mild in flavour, and of poor keeping quality.

Spring onions (4) are used mainly in salads. They are always grown from seed and harvested when young. 'White Lisbon' (4) is the most popular variety for this purpose.

5 Chives (*Allium schoenoprasum*). As a native plant, chives are widespread in the northern hemisphere, from arctic Russia to Japan, from northern Europe to parts of the Mediterranean region, and in North America. They occur also in Asia Minor and the Himalayas. In Britain, they are found wild in some rocky pastures and occasionally become established in other places where they have escaped from cultivation. Chives grow in tufts, producing bright green, slender, cylindrical leaves, 3 to 9 inches long, and taller flowering stems bearing terminal, globose heads of pale purple or pink flowers. They do not have swollen bulbs and so the only part used is their mild-flavoured leaves, which are usually chopped and used for garnishing, or in soups, omelettes, salads, and sandwiches. They are primarily home-grown, ornamental as well as useful in gardens and window-boxes, but there is a limited market for them, mainly for hotel and restaurant use.

6 Welsh Onion (*Allium fistulosum*). Despite its name, the Welsh Onion is not a native of Wales nor has it ever been cultivated extensively in that country. The word may be a corruption of the German 'welsche' (foreign), applied when this onion was introduced into Europe towards the end of the Middle Ages. Other names for this species are 'Japanese Bunching Onion' and 'Ciboule'. It is not known in the wild state but it probably originated in eastern Asia, the home of the closely related *Allium altaicum*. It has been the principal garden onion in China and Japan since prehistoric times but has never become popular in the west.

The specific epithet, 'fistulosum', refers to the fistulose or hollow, cylindrical leaves and flowering stem, the latter about 20 inches high and bearing a globose head of yellowish-white flowers. The bulbs of this onion are elongated and only slightly swollen. In Britain, *Allium fistulosum* is regarded as a home-garden crop, used as a substitute for spring onions and for seasoning.

1 ONION *diagram* STEM

1A

4

3

5   2

6

TWO-THIRDS LIFE SIZE   *ONION PLANT* × $\frac{1}{8}$

1 ONION *diagram*   1A Plant

2 'SPANISH' ONION   3 ONION 'BLOOD RED'   4 SPRING ONION 'WHITE LISBON'

5 CHIVES   6 WELSH ONION

1 **Shallots** were long thought to be a distinct species, and masqueraded under the name '*Allium ascalonicum*'. They are nowadays considered to be merely a variety of onion, belonging to the Aggregatum group. This group, which also includes the 'potato onion' or 'multiplier onion', differs from the common onion group in that its bulbs multiply freely, producing several lateral bulbs (1). Propagation is usually vegetative, by dividing the cluster of bulbs and replanting them singly, but seed may be produced by some strains. Some shallots, e.g. 'Jersey Shallot', tend to produce side bulbs and develop divided crowns, like the potato onion. Shallots are mainly a garden crop, though there is some commercial production for pickling. The proverbial instruction for shallots — 'Plant on the shortest day, lift on the longest day' — is a useful guide.

Tree Onion, Egyptian Onion, Catawissa Onion. These forms all belong to the Proliferum group of onions, in which the inflorescence produces a cluster of bulbils or small bulbs instead of seed. Sometimes both bulbils and flowers are borne in the same inflorescence. They are of little commercial value, the bulbils often being too small to be of much use.

2 **Leek** (*Allium ampeloprasum* var. *porrum*). The modern garden leek is not known in the wild state, but it shows a distinct affinity with the Wild Leek (*Allium ampeloprasum*) which is native in the Mediterranean region and Macaronesia (that is, the Atlantic islands of the Azores, Canaries, Cape Verdes, and Madeira), and possibly native in the few areas on the southern coasts and off-shore islands of England and Wales where it occurs. Leeks probably originated in the Eastern Mediterranean region or the Near East, where they have been cultivated for about three or four thousand years. The leek was known in Europe during the Middle Ages and today is an important crop in many countries, including France, Britain, northern Spain, Belgium, Denmark, and Holland. Leeks are grown from seed; they may be sown under glass in January or February for planting in the open in spring or early summer; or they may be sown in the open in March. To produce high quality leeks, it is necessary to blanch the lower part of the plant. This is done by planting the seedlings in trenches or furrows and drawing the soil up to exclude light. The blanched, elongated bulb is the part the cook uses — in soups and stews, or as a separate vegetable, boiled, or par-boiled and fried in batter.
The Welsh traditionally wear a leek on St. David's Day to commemorate the victory of King Cadwallader over the Saxons in 640 A.D.; before this battle the Welsh gathered leeks from a nearby garden and wore them in their hats as a distinguishing mark, thereby avoiding mistaken identity and accidental attacks on their own men.

The leek has flat leaves folded sharply lengthwise, their long bases encircling each other to form an elongated, cylindrical bulb (2c). The flowering stem is stout, cylindrical, terminating in a more or less globose inflorescence (2A) which is at first enclosed within a papery spathe with a long beak. The flowers are pale purple, with exserted stamens (2B).

Kurrat. This plant is closely related to the leek, and similar to it but smaller, with narrower leaves. It probably originated in the eastern Mediterranean region, and is grown there, especially in Egypt, for its edible tops.

3 **Garlic** (*Allium sativum*). Believed to be a cultivated race derived from a plant which grows wild in central Asia, garlic was popular in ancient Egypt, but the history of its introduction into the British Isles is obscure. Because of its pungency, it is generally regarded as a flavouring agent rather than as a vegetable in its own right, and its flavour is less appreciated in the British Isles than in southern Europe. This is partly due to its misuse. For most purposes, it should be crushed very finely and used in moderation. If fried in fat or oil which is too hot, it develops an acrid flavour.
Garlic cloves, separated by breaking apart the parent bulb, are planted in autumn, winter, or very early spring, about 1½ inches deep and 6 to 8 inches apart, in rows. Unlike most other edible Alliums, garlic bulbs develop entirely underground (3). They are ready for lifting when the leaves wither, usually in July. Garlic bulbs are made up of several cloves enclosed within the white or pink 'skin' of the parent bulb (3B). The leaves are flat and rather slender, the flowering stem smooth and solid. The whitish flowers are usually mixed with bulbils. The young flower-head is enclosed in a papery, long-beaked spathe, which is soon shed (3A).

Rocambole. This name should be used for forms of *Allium sativum* which have coiled stems, but it has often been applied to an allied species, *Allium scorodoprasum*, also called 'Sand Leek', a very local native plant which is occasionally cultivated and used for the same purposes as garlic.

Chinese Chives (*Allium tuberosum*) are little known in the West, but an important vegetable in parts of eastern Asia, from Mongolia to the Philippines. They have conspicuous rhizomes but little-developed bulbs. The parts used are the young leaves and flower-stalks. They have a strong flavour, resembling that of garlic, and are used for seasoning meat dishes.

PLANTS × ⅛   *INFLORESCENCES AND BULBS* × ⅔   *DETAILS* × 3

1 SHALLOT plant   1A Inflorescence   1B Flower detail   1C Bulbs
2 LEEK plant   2A Inflorescence   2B Flower detail   2C Elongated bulb
3 GARLIC plants   3A Inflorescence   3B Composite bulb

SALAD ROOTS

1 Beetroot (*Beta vulgaris*). The Beetroot is closely related to the sugar beets, mangolds, spinach beets, and chards, which are believed to be derived from the wild subspecies *maritima*, native on sea-shores in Europe (including the British Isles), in North Africa, in the Azores, and from Asia Minor to the East Indies. Beetroot is commonly used as a salad root, boiled in water, peeled, and eaten cold, either freshly cooked or pickled in vinegar. Boiled beetroots are also eaten as a hot vegetable, either plain or prepared in various ways — fried, or served with various sauces, or hollowed out and stuffed with savoury mixtures. *Borshch* is a famous Russian soup, in which one of the principal ingredients is the red purée of cooked beetroot. Beetroot wine is a popular home-made wine. Being closely related to sugar beet, it is not surprising that the garden beetroot has a high sugar content.

Horticulturally, beetroot can be classed in three categories: 'globe' (illustrated), spherical, with a small tap-root — for example 'Detroit Red Globe' grown especially for early crops and for canning; 'long' — for example, 'Cheltenham Green Top', with larger, tapering roots; and 'intermediate', — for example, 'Feltham Intermediate', with oblong ovoid roots.

Beetroot belongs to the Goosefoot family (Chenopodiaceae). It is annual or biennial in garden forms, but the wild subspecies, which has a stout but not fleshy tap-root, is perennial. The leaves are variable in shape and colour, but often rhomboid-ovoid near the base of the plant and dark green or reddish. The small, green, hermaphrodite flowers are borne in cymes, arranged in a tall, branching, spike-like inflorescence. The fruits are 1-seeded, but 2 or more are usually joined together by the swollen perianth bases to form a 'seedball' which when sown can be expected to produce more than one seedling.

2 Radish (*Raphanus sativus*). Unknown in the wild state and of ancient and uncertain origin, radishes were grown in Egypt over 2,000 years ago and various forms are cultivated in most parts of the world. In Britain, the roots of radishes are usually eaten raw, in salads. Those with red or red and white roots are the most popular, such as the (illustrated, from left to right) round, red 'Scarlet Globe'; the cylindrical, red and white 'French Breakfast'; and the round, red and white 'Sparkler'. White radishes, such as 'Icicle', are less frequently grown. Salad radishes are available practically the whole year round, from sowings outdoor in spring and summer, and in greenhouses and frames in autumn, winter, and early spring. They grow very quickly when conditions are suitable and are usually ready for picking within 3 or 4 weeks. Warm, light, sandy but highly fertile soils are the most satisfactory, and an adequate water supply is essential.

3-4 Winter Radishes are so-called because their solid firm-fleshed roots, if lifted in autumn, can be stored for winter use without becoming hollow in the centre — as salad radishes would if they were allowed to grow longer than usual and then kept in storage. Winter radishes, which are much larger than the more familiar kinds, are usually sown in summer and allowed to grow for several months before being lifted. They are not commonly grown in Britain, but two are often listed by British seed merchants: 'Round Black Spanish' (3) which has turnip-shaped, dark-brown roots, quite unlike most other radishes in appearance; and 'China Rose' (4) which has large, cylindrical, bright red roots.

Radishes belong to the Wallflower family (Cruciferae) and have characteristic cross-shaped flowers with 4 green sepals and 4 white, lilac, or pinkish petals. Their fruit is a rather fleshy siliqua, 1 to 3 inches long, with a pointed beak. Their leaves are lobed and irregularly toothed, with a large terminal segment and smaller, paired, lower segments.

The **Rat-tailed Radish** (*Raphanus caudatus*) is grown in southern Asia. The part eaten is not the root but the fruit, which reaches a length of 8 to 10 inches. It has a somewhat pungent flavour, similar to that of a salad radish root, and is eaten raw or pickled in vinegar.

CRUCIFER AND COMPOSITE ROOT CROPS

1 **Turnip** (*Brassica rapa*) has been known in Europe from prehistoric times and its cultivation has spread into most parts of the world. It is grown primarily for its swollen root, eaten cooked in various ways, but the green leafy tops are also eaten as 'spring greens'. Turnips are also used as fodder for farm livestock, large, coarse roots being preferred for this purpose.
The turnip belongs to the Wallflower family (Cruciferae). It is a biennial. Roots which are left in the ground over the winter produce a crop of 'turnip tops' which can be used as spring greens in March. The swollen turnip 'root' consists mainly of the hypocotyl — that portion of the axis which lies between the true root and the first seedling leaves (cotyledons). This distinguishes it from the swede, in which both the hypocotyl and the base of the leafy stems are swollen. In flower, the turnip may be distinguished from either swedes or cabbages by the way in which its open flowers become raised above the unopened buds (1A). Their 4 petals are bright yellow. The cultivated turnips vary considerably in the size, shape and colour of their 'roots'. They may be round, flattened or cylindrical, yellow or white, with or without a green or purple zone near the top. Early sowings are made in hot beds and cold frames for succession. Summer and winter crops are sown from March to mid-July.

2 **Swede** (*Brassica napus* var. *napobrassica*). This is a comparatively recent vegetable, believed to have originated from a hybrid between *Brassica oleracea* and *Brassica rapa*, probably in Bohemia and possibly as recently as the 17th century, for there are no earlier records. Swedes are grown both for human food and for feeding to livestock. The large fodder varieties are quite suitable for human consumption and are preferred by some to the smaller and often milder-flavoured garden varieties. Swedes are used mainly in stews, or served 'mashed' as a separate vegetable. As the swollen 'root' is composed of both the hypocotyl and the base of the leafy stem, a swede can be distinguished from a turnip by the presence of a swollen 'neck' bearing a number of ridges, the leaf-base scars.

Swedes may be purple, white, or yellow, with yellow or, less commonly, whitish flesh. The plant is usually a biennial. When in flower, it may be distinguished from a turnip by its open flowers (2A) not being raised above the unopened buds.
Seed is usually sown in May or June. The crop is ready for lifting in late summer, but it is hardy enough to be left in the ground all winter if need be.

3 **Scorzonera** (*Scorzonera hispanica*). The fleshy roots, black-skinned but with white flesh, are eaten as a boiled vegetable, like salsify (the plant is sometimes called 'Black Salsify'). They contain inulin (a sugar which can be eaten by diabetics), and have a distinctive sweet flavour. Another use is as a coffee substitute, in the same way as chicory (*see* p. 110). The young leaves are sometimes eaten in salads.
Scorzonera belongs to the Daisy family (Compositae). It is a native of central and southern Europe. Although naturally a hardy perennial, it is usually grown as an annual or biennial. Seed is sown in spring, in rows, and the seedlings are thinned out to about 9 inches apart. The cylindrical tap-root is ready for lifting in autumn, but will continue to increase in size if left in the ground for another year. The dandelion-like, yellow flower-heads (3A) are borne on long stalks, on flowering stems 2 to 3 feet high. The 'seed' bears a pappus of feathery hairs.

4 **Salsify** (*Tragopogon porrifolius*) is also called 'Vegetable Oyster' or 'Oyster Plant', because its root is supposed to taste like oysters. The white roots are eaten throughout the winter, boiled, baked, or as a creamed soup, and the tender young leaves make a palatable salad.
Salsify belongs to the same family as Scorzonera and has similar cultural requirements. It too is a native of southern Europe. In the British Isles, it sometimes escapes from gardens. The plant is usually a biennial, 2 to 3 feet high, with terminal, long-stalked, purplish flower-heads (4A). The 'seed' bears a pappus of both feathery and simple hairs.

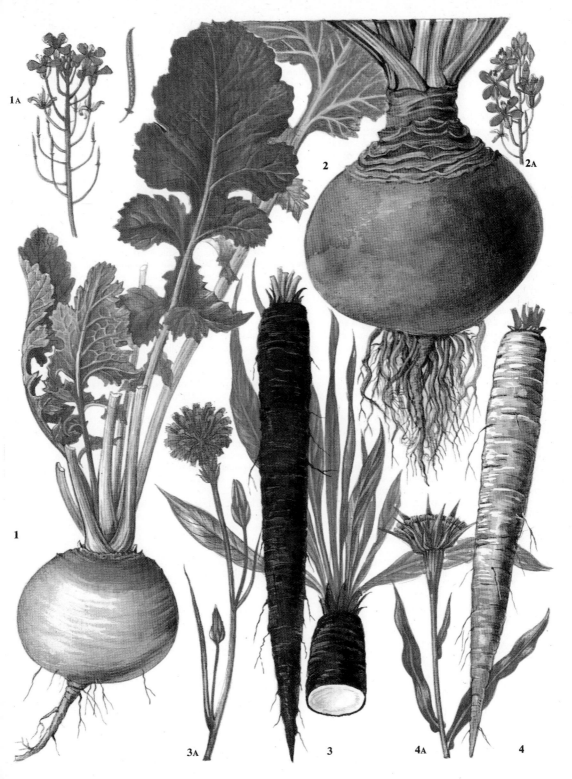

HALF LIFE SIZE

1 TURNIP   1A Flowers and fruit   2 GARDEN SWEDE   2A Flowers
3 SCORZONERA   3A Flower-head   4 SALSIFY   4A Flower-head

THE UMBELLIFER ROOT CROPS

1 Carrot (*Daucus carota*). The wild carrot, which grows in grassy places, especially near the sea, is a biennial herb with a comparatively small, tough, pale-fleshed tap-root bearing little resemblance to the thick, fleshy, orange or reddish tap-root of the cultivated carrots, which many botanists regard as being a distinct subspecies, ssp. *sativus*. Carrots were known to the Greeks and the Romans and are now cultivated all over the world. They are a rich source of vitamin A and their high sugar content is indicated by their sweetness. Besides being one of the most valuable root crops grown for human consumption, they are useful for feeding livestock. Intermediate or half-long cultivars (1B), e.g. 'James's Intermediate', are the kind usually grown in the open for the main crop; but 'Chantenay Red Cored' may be grown either in the open or under glass. Stump-rooted carrots (1D) may be used for forcing, or for growing on heavy land, but they are less popular than the longer kind. 'Amsterdam' is a half-long, cylindrical carrot (1C), much used for forcing.

The finely divided leaves make the carrot plant quite ornamental, and if left to grow a second year it bears a terminal compound umbel of white flowers (1A), subtended by ternate or pinnatifid bracts, the central flower of each umbel often red or purple.

2 Parsnip (*Pastinaca sativa*). Wild parsnip is locally abundant in England and Wales and also occurs in much of Europe and the Caucasus. It has been introduced into other parts of the world. Parsnips have been cultivated at least since Roman times, but good, fleshy forms do not seem to have been developed before the Middle Ages. Their fleshy roots (2B) contain quantities of sugar and starch and are used as food for man and livestock and also for making parsnip wine. Cultivars have larger, fleshier roots and longer leaves than the wild plants, but they are basically very similar. In Scotland and Ireland, apparently 'wild' plants have escaped from cultivation. The parsnip is recognizable by its characteristic smell, hollow, furrowed stems, and large, simply pinnate leaves with ovate, lobed and toothed leaflets. The small, yellow flowers (2A) are borne in an umbel up to 4 inches across.

3 Celeriac (*Apium graveolens* var. *rapaceum*) is closely related to celery (*see* p. 149) but whereas the stem is the edible part of celery, the part of the celeriac that is eaten is the swollen base of the stem. It is sometimes called 'Turnip-rooted celery', being superficially of turnip-like appearance, although not a root crop in the botanical sense of the term. Unlike celery, celeriac leaf-stalks are little swollen and of moderate length; also, they are bitter and therefore unfit for use in salads. The swollen 'root' is usually rather irregularly sub-globose in shape, about 4 inches across, brown outside, with whitish flesh. It may be eaten boiled, as a separate vegetable, in soups, stews, etc., or grated, in salads. The flavour is similar to that of celery.

Celeriac has never equalled celery in wide popularity but it is grown commercially in Britain on a limited scale, and seed is available from most seed merchants.

Parsley, Turnip-rooted (*Petroselinum crispum* var. *tuberosum*). The culinary part of this plant is the thick, fleshy, dingy-white root, rather like a small parsnip in appearance but with a flavour which has been compared with that of celeriac. The leaves are compound, as in wild parsley, not curled and crisped like most of the cultivars grown for their leaves. Turnip-rooted parsley is not an ancient vegetable but the history of its development in cultivation is not known. It has never achieved much popularity in this country, but — as one of its names 'Hamburgh Parsley' suggests — has been grown more extensively in Germany.

Chervil, Turnip-rooted (*Chaerophyllum bulbosum*). This biennial is a native of southern Europe, sometimes cultivated for its roots, which are 4 or 5 inches long, grey or blackish, with sweet, yellowish flesh. They are lifted in summer, when the foliage dies back, and may be stored like potatoes. They are used in stews, or boiled and served as a separate vegetable.

PLANTS × ⅛   *FLOWER-HEADS AND ROOTS* × ½
1 CARROT plant   1A Flower-head
1B, 1C, 1D Roots of intermediate main-crops, forcing, and stump-rooted varieties
2 PARSNIP plant   2A Flower-head   2B Mature plant   3 CELERIAC plant   3A Root

175

POTATOES

Today, the potato is one of the most important food plants of the world. The 'Irish' or 'European' Potato (*Solanum tuberosum*) is descended from plants which originated in the temperate Andes of South America where, it is believed, potatoes have been used for food for some 2,000 years. As a crop of world-wide importance, the potato is a relative newcomer. Although the Spanish conquistadores recorded that it was cooked and eaten by the South American people, they do not seem to have rated it very high. Despite this, the potato was probably first introduced into Europe by way of Spain, during the latter half of the 16th century, perhaps as cook's stores on ships sailing from the Colombian port of Cartagena. The stories that attribute its introduction to Sir Walter Raleigh or Sir Francis Drake seem to be mere legends, although the evidence is inconclusive. The potatoes of the 16th century apparently belonged to the *andigenum* series, producing late-maturing tubers, narrow in proportion to their length, and very 'knobbly', with deep eyes. The earlier-maturing, heavier-cropping, relatively smooth and shallow-eyed modern potatoes are believed to have been derived from the *andigenum* series by selection, both in South America and in Europe.

The Irish were the first people in Europe to make great use of the potato in the early 17th century. (Their economy became so dependent on the potato that the failure of the crop in 1845 brought about a famine which started the wholesale emigration of the Irish people which has continued to some extent even to the present day.) The crop was not widely planted in other parts of the British Isles or in other European countries until several decades later, and it did not become a really important item in the diet of the English working classes till the late 18th century. The major producing countries now are in the temperate zones, but potatoes can be grown almost everywhere except in low tropical regions. They are cultivated up to altitudes of about 8,000 feet, and as far north as central Alaska, southern Greenland, and northern Scandinavia.

Under suitable conditions, potatoes yield a higher food-value per acre than any cereal. Their principal disadvantages, in comparison with cereals, are their high water content — which adds to the relative cost of transport — and their shorter storage-life and greater wastage. As a freshly cooked vegetable they may be served in a great variety of ways — boiled, steamed, fried, baked, roasted, or as an ingredient of soups, stews, pies and rissoles. They are also processed for sale as potato crisps, potato flour, and dried mashed potato. Potato starch, and dextrose produced by hydrolysis of starch, also have their uses in the food industries. Alcohol prepared by the fermentation of cooked potatoes is an important product in some countries for the manufacture of liquor, such as schnapps. Considerable quantities of potatoes are used for feeding livestock.

For agricultural or horticultural purposes, potatoes are classified as First Early, Second Early, or Maincrop, according to the date when they are normally marketed. They are also distinguished by differences in shape and skin-colour — those with a brown skin being oddly termed 'white'. About 80 cultivars are listed in the seed growers' registers in this country, but 75 per cent of the total acreage in England and Wales is occupied by only three cultivars — 'Arran Pilot', 'Majestic', and 'King Edward'. These and the other important cultivars illustrated are classified as follows:

2 'ARRAN PILOT' — First Early, white, kidney.
3 'KING EDWARD' — Maincrop, part pink, oval.
4 'MAJESTIC' — Maincrop, white, kidney.
5 'CRAIG'S ROYAL' — Second Early, part pink, oval.
6 'RED CRAIG'S ROYAL' — Second Early, pink, oval.
7 'HOME GUARD' — First Early, white, oval.
8 'RECORD' — Maincrop, white, round.
9 'DR. McINTOSH' — Maincrop, white, long oval.

The potato is a perennial herb of the Solanaceae family, with rather weak, straggling or more or less erect, branching stems, 1 to 3 feet high. It has odd-pinnate leaves with 3 or 4 pairs of ovate leaflets, with smaller ones in between. The flowers (1A) are white to purplish, about 1 inch across, with yellow anthers joined laterally to form a cone-shaped structure which conceals the ovary. The fruit (infrequently produced) is a tomato-like green or yellowish berry, about $\frac{3}{4}$ inch across. The plant has fibrous roots and many rhizomes (underground stems) which become swollen at the tip to form the edible tubers which are the potatoes we eat. The haulms (stems and leaves) die down in autumn, leaving the tubers to survive the winter underground. In practice, they are usually lifted and stored in 'clamps', protected with straw. The fresh tubers which are planted in spring are usually obtained from certified disease-free stocks grown in favourable areas, particularly Scotland and Ireland. Potatoes used for planting are called 'seed potatoes', although botanically they are tubers, not seeds. The young shoots develop from the 'eyes' of the old potato, taking nourishment from it, so that it slowly withers and dies.

Potatoes are a valuable article of diet, containing, besides almost 78 per cent water, 18 per cent carbohydrates (mainly starch but also a little sugar), 2 per cent proteins, and some potash. All green parts of the plant (including 'potatoes' which have been exposed to light) contain poisonous alkaloids (solanines).

TUBERS × ⅔　　　　PLANT × ⅛　　　　FLOWER DETAIL × 1

1 POTATO plant　　　1A Flower detail　　　1B Seed potato
2 'ARRAN PILOT'　　　3 'KING EDWARD'　　　4 'MAJESTIC'　　　5 'CRAIG'S ROYAL'
6 'RED CRAIG'S ROYAL'　　　7 'HOME GUARD'　　　8 'RECORD'　　　9 'DR. McINTOSH'

OTHER TUBERS

1 **Jerusalem Artichoke** (*Helianthus tuberosus*) has crisp-fleshed, underground stem-tubers (1A), whitish or yellowish in colour, sometimes tinged with pink, irregularly club-shaped, oval or roundish, usually with short, knobbly branches. Selected varieties are less irregular in form and up to 4 inches long and 2 – 3 inches across. The tubers can be eaten boiled, baked, in stews, soups, etc. They are sweet-fleshed, due to the presence of inulin, a sugar which can be eaten by diabetics.

The Jerusalem Artichoke is a member of the Daisy family (Compositae) and closely related to the Sunflower. Both are natives of North America. The tubers are usually planted in early spring and lifted when needed in autumn and winter. The stout stems, up to about 6 feet high, bear large, rough-surfaced leaves. In Britain the sunflower-like, yellow flower-heads are borne only after a long, hot summer.

2 **Oca** (*Oxalis tuberosa*). Though long cultivated in its native Peru, Ecuador, and Bolivia, oca is an uncommon vegetable in European gardens. Its white, yellow or red tubers (2A) are cylindrical, with a series of grooves and bulges. They are very acid when fresh and in South America are dried in the sun for a few days, when they become floury and less acid. If dried for weeks, they are said to acquire a flavour similar to dried figs. Oca tubers are not hardy and are sometimes started into growth under glass, to be planted out in May. The leaves and young shoots may be eaten in salads or cooked, like sorrel.

Oxalis is a large genus of 500 or more species, which gives its name to the Wood Sorrell family (Oxalidaceae).

3 **Ulluco** (*Ullucus tuberosus*) is a vegetable which is little known outside its native western South America, although the Horticultural Society of London grew it more than 100 years ago. It is a perennial herb. Its creeping, pink stems root where they touch the ground and produce slender rhizomes, which become swollen at their tips to form small, underground pink or yellow tubers. The flesh is mucilaginous and starchy.

Ulluco belongs to the Basella family (Basellaceae), a small family found mostly in tropical America. It is half-hardy and grows best in a light, rich soil, with plenty of leaf-mould, the tubers being planted in spring and lifted in October or November, after frost has killed the foliage.

4 **Ysaño** (*Tropaeolum tuberosum*) is also known as 'Tropaeolum' or 'Tuberous Nasturtium'. It is a perennial climber, with herbaceous stems up to 5 or 6 feet high. Its tubers, yellow or greenish, marbled with purple markings, are eaten in Peru, Chile, and Bolivia. If simply boiled, like potatoes, their flavour is said to be disagreeable. In Bolivia, they are considered to be a delicacy if frozen after boiling, and in other areas they are partially dried before being eaten.

The genus Tropaeolum, which includes several ornamental garden plants, gives its name to the family Tropaeolaceae. This particular species, although less spectacular than some, is worth growing as an ornamental climber. Its yellow and red flowers, although smaller than those of the more familiar garden nasturtiums, are held well above the leaves on long red stalks. The plant is almost hardy and is easily propagated by planting the tubers in spring. They are not ready for lifting until late autumn.

1 **2** **2A** **3** **1A** **4** **4A**

TWO-THIRDS LIFE SIZE   *FIG.* $1 \times \frac{1}{12}$

1 JERUSALEM ARTICHOKE   1A Tuber   2 OCA   2A Tubers
3 ULLUCO   4 YSAÑO (TROPAEOLUM)   4A Tubers

179

STARCHY ROOTED PLANTS USED IN MAKING
ARROWROOT AND TAPIOCA

1 **Cassava** (*Manihot utilissima*). In Asia this plant is usually called 'tapioca', a name which in other countries is reserved for the manufactured product prepared from the roots; in some countries it is called 'manioc'. Originally a native of America, it is now one of the most important food plants in all parts of the tropics (except at the highest altitudes) where there is a heavy rainfall. The plant is a shrub growing to 6 – 7 feet or more. It has woody stems, sections of which are used as cuttings to propagate the crop, and swollen tuberous roots (1A), which are the part eaten. Harvesting of different varieties takes place at from about 8 months to 2 years after planting. The varieties are conventionally divided into 'sweet' and 'bitter'. The bitter varieties have a higher content of certain substances from which prussic acid can arise, and are extremely poisonous to people or livestock if eaten raw. These varieties have to be particularly carefully prepared for eating, for example by prolonged or repeated boiling. The characteristics of sweetness and bitterness may, however, change within a variety under differing conditions of soil and climate. The most serious disease of the crop is cassava mosaic, caused by a virus, which occurs only in Africa but often severely limits yield there.

Cassava has the great virtue that many varieties, after reaching maturity, can be left in the ground for another two years or so without weeding or other attention and without deterioration of the tubers. The crop is almost immune to locust attack. For these reasons, a plot of cassava in the ground is as typical a famine reserve of the wet tropics as is a granary of millet in the dry tropics. But cassava also has a disadvantage. Its nutrient content consists almost entirely of starch, with a particularly low protein content — so low that a prolonged diet depending too heavily on cassava can lead to serious malnutrition.

Cassava hardly enters into world trade, but is consumed almost entirely where it is grown. The most usual form of preparation is to remove the peel from the roots (1A) and then boil and mash them. In some countries the roots are grated to produce a meal, known as 'farinha' in Brazil and 'garri' in Nigeria, which can be cooked in small cakes and is a convenient travellers' food; the meal is sometimes lightly fermented. The juice of the bitter varieties can be boiled down to make 'cassareep', which is used in some West Indian sauces. Manufactured tapioca is prepared, for example in Chinese factories in Malaysia, by washing out the starch from the tubers and drying it in either the pearl or flake form for export. This product is mostly used in temperate countries in puddings, biscuits, and confectionery. An alcoholic liquor can be prepared by fermenting cassava, and the leaves of the plant are sometimes boiled and eaten as a vegetable.

2 **Arrowroot** (*Maranta arundinacea*) is a tropical herbaceous perennial with swollen starchy rhizomes (2A). The starch is in very fine grains which are easily digestible and is thus particularly suitable for invalid diets. Almost the whole of the world export supply is provided by the island of St. Vincent in the West Indies. The plants are propagated by rhizomes or suckers, and dug up about 10 – 11 months later. The starch is prepared from the rhizomes by peeling, washing, grating, and sifting in repeated operations followed by drying. The limited demand for arrowroot relegates the crop to a minor status.

3 **Taro** (*Colocasia antiquorum*) is known in the Pacific islands as 'taro'; in the West Indies it is called 'eddo' or 'dasheen', and in West Africa 'old cocoyam'. It is a root crop of secondary importance grown in many parts of the wet tropics, and does best on a moist or even slightly swampy soil. The part eaten is the corm (3A) which is formed underground by a thickening of the base of the stem; some varieties produce also subsidiary tubers called 'cormels' (as shown in 3). The crop is propagated by cutting off the top part of the corm and planting it, or by planting cormels. The plants grow 2 – 3 feet high, rarely flower, and produce very large leaves of the same shape as the Arum lily, to which the species is related; the leaf-stalk is attached near the centre of the leaf-blade. Taros are consumed locally, mostly by those who grow them; the starch is, as in arrowroot, very fine-grained and easily digestible.

4 **Tannia** (*Xanthosoma sagittifolium*), is also sometimes known as 'yautia' and in West Africa as 'new cocoyam'. It is of the same family as taro but can be distinguished from it by the leaf-stalk being attached to the edge of the leaf at the apex of a notch, which makes the leaf arrow-shaped. The plants are also taller than the taro (up to 7 feet) and are usually planted on drier ground, though the crop is grown in the same wet tropical regions. Propagation is by the same methods as for the taro. In this crop the main corm always produces a number of subsidiary tubers.

Other plants used for Arrowroot. A number of plants grown in different parts of the tropics and having starchy edible roots are sometimes described as kinds of arrowroot. 'Queensland arrowroot' (*Canna edulis*) has rhizomes of this kind and the tops are also sometimes used as cattle fodder. 'Indian arrowroot' is *Curcuma angustifolia* and 'African arrowroot' is *Tacca involucrata*. But these and some other similar plants are only of very local and minor importance.

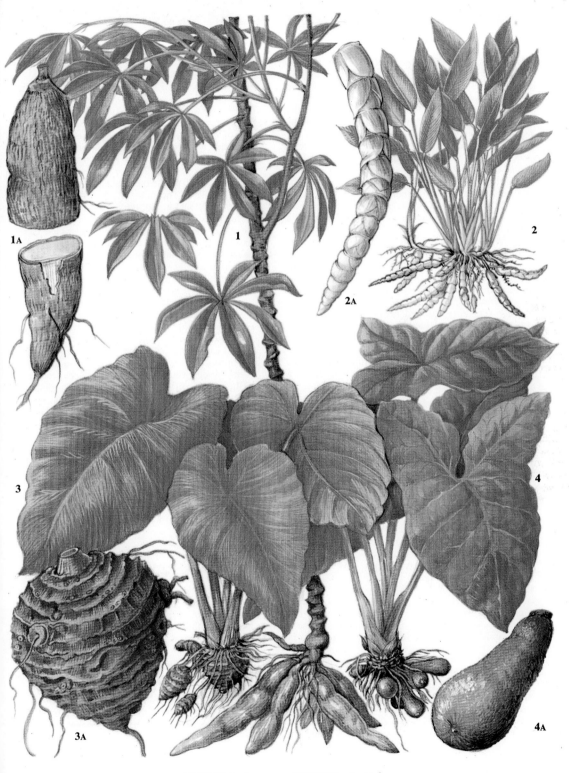

1 CASSAVA   1A Root   2 ARROWROOT   2A Rhizome
3 TARO   3A Corm   4 TANNIA   4A Tuber

SWEET POTATOES AND YAMS

1 Sweet Potatoes (*Ipomoea batatas*) are not closely related botanically to ordinary potatoes, and the chief resemblance between the two crops is that both have edible tubers of about the same size. Originating in South America, sweet potatoes are now commonly grown all over the wetter tropical regions, usually as a secondary rather than a staple food, with the highest consumption perhaps in some islands of the Pacific. The crop is also grown in the warm temperate zone, as far north as Spain and as New Jersey in the United States. In the tropics, it is propagated by planting leafy stem cuttings known as slips. In the temperate zone, where the plants do not over-winter in the field, this is not possible; so tubers are grown under forcing conditions in the spring to produce a crop of shoots which can be planted out. The crop takes from about 4 to 8 months in different varieties to reach maturity. The haulm or vine above ground produces a dense mass of foliage, and in the tropics the plants often bear flowers (1c) though they rarely set seed; this haulm is a good fodder for livestock, and is sometimes cooked and eaten by people in times of famine. The tubers formed below ground are most commonly of an elongated shape (1A, 1C) but sometimes more nearly spherical (1, 1B). The outer skin may, as shown, be either white or a shade of red or purple. The inner flesh is usually white, but in some varieties yellow. (The latter are especially useful as a source of vitamin A in the diet.) The nutritional value of the sweet potato is chiefly as a source of starch; there is a small amount of protein, and some sugar content which gives the tubers their slightly sweet taste. Sweet potatoes are usually cooked by boiling and mashing. Their dry matter content is higher than for ordinary potatoes, so that a somewhat smaller helping will be adequate. They do not store well, and in temperate countries sweet potato tubers in store must be carefully protected against frost.

2 Yam. The word 'yam' describes the cultivated species of the genus *Dioscorea*, but confusion can arise as the name is sometimes loosely applied to almost any tropical root crop, and in America to sweet potatoes. About 10 species of yam provide important food supplies; all are of Old World origin except the cush-cush yam *Dioscorea trifida* which is native to America. Most species are essentially tropical, but *D. opposita* and *D. japonica* are cultivated in northern China and Japan, and the former has been grown in Europe.

Yams produce either one large tuber (2B) or in a few species, such as the Chinese and cush-cush yams, a number of smaller ones. In the latter case, a single whole tuber is used for planting; in the former, the top or 'crown' of the tuber, which contains buds or eyes, is cut off and used for planting. Like many other root crops, yams are commonly planted on mounds or ridges. As they are climbing plants, they are generally provided with stakes or other supports to grow up; a difference between species is that some twine clockwise and some anti-clockwise. The period from planting to harvest is about 8–12 months. There are two areas of the tropics where yams are most important in the diet. One is the high-rainfall zone of West Africa from the Ivory Coast to Cameroon. Here the most important species are *D. rotundata* the white yam, which is the one illustrated (2, 2A), and *D. cayenensis*, the yellow or Guinea yam. In the former the flesh of the tuber is white, and in the latter usually yellow. The other main area of production centres on Vietnam, Cambodia, and Laos and extends to surrounding countries. Here the most important species is the Asiatic yam, *D. alata*, which can be readily recognised by the winged petioles of the leaves. The most important secondary species, in this and several other areas, is the Chinese yam, *D. esculenta*. Besides these two main regions, there are others where yams are grown on an important scale, especially the West Indies and the Pacific islands. A peculiar species of yam, *D. bulbifera* has root tubers which are not very good to eat but bears aerial tubers on the stem, which are the part eaten. There are species of wild yam, growing in both Africa and Asia, whose tubers are collected for eating in times of food shortage. Yams, after peeling the tuber, can be cooked in various ways, of which the West African method of boiling and mashing is the commonest; but roasting and frying are also widely used. In some West African areas, for example in south-eastern Nigeria, the consumption of yams is so high that they must be regarded as the staple food of the population. Since yams contribute little but starch for human nutrition, there is a risk of protein deficiency among populations using such a diet. Yams store better than most tropical root crops, and can be kept for some months by the West African method of placing them on racks or hung by strings in an open-sided shed.

3 Yam Bean. This name is applied to several species of the family Leguminosae which produce both edible tubers and edible seeds or pods. The most important is *Pachyrhizus erosus* which is native to the American tropics and also grown as a minor crop in south-east Asia. The forms grown in Asia usually produce a single smoothly-rounded tuber, but the tubers may also be lobed and have subsidiary tubers (as shown in 3). The flesh of the tuber is white and chiefly supplies carbohydrates; its protein content, although the plant belongs to the Leguminosae, is small. The pods of this species (3A) are eaten whole and only when young, as in the mature state they can be poisonous. The other best known yam bean is *Sphenostylis stenocarpa*, which is occasionally cultivated in West Africa.

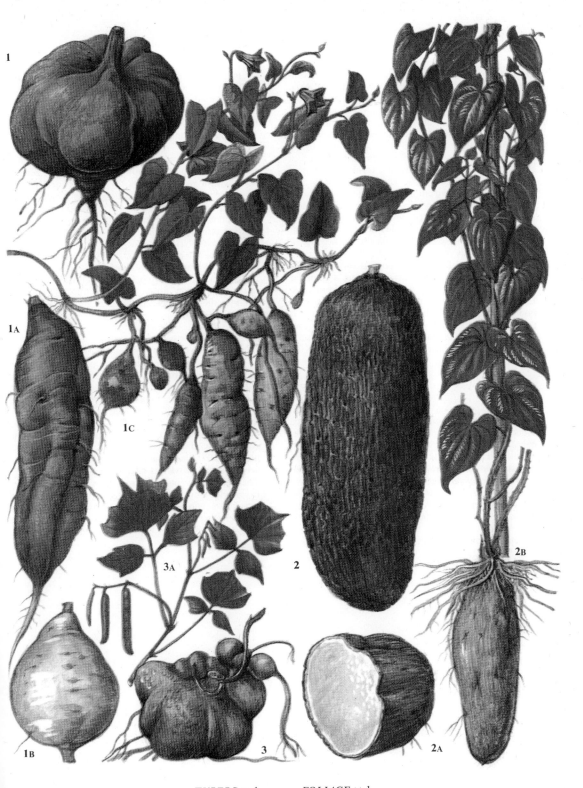

TUBERS × ¼   *FOLIAGE* × ⅛

1, 1A, 1B SWEET POTATO tubers   1C Flowering plant with tubers
2, 2A YAM tubers   2B Plant with tuber
3 YAM BEAN tuber   3A Bean bearing shoot

183

SAGO AND SUGAR PALMS

1 **Sago Palm** (*Metroxylon sagu*). This tree, which is the most important source of sago, grows wild in fresh-water swamps in the countries of south-east Asia. It can be propagated by suckers, but there is often little need for this as the palms in many sago swamps are self-regenerating and the only attention needed may be judicious thinning if they are too close. The palm only flowers (1A) once in its life, at the age of about 15 years. Just before this there is a build-up of starch reserves in the pith of the trunk, and it is at this stage that the trees are felled. The bark is then removed and the stem cut into sections a few feet long (as shown at the foot of Fig. 1). These logs are next split and the pith scooped out. Rasping or grinding of the pith is followed by washing out and settling off the starch which if dried at this stage forms sago flour, much of which is consumed locally. To prepare pearl sago, the wet starchy paste is pressed through a fine sieve and dried on a hot surface, giving a granular product. It is this form which is most familiar in European cookery, where it is used chiefly in making puddings and sweet dishes. The food value of sago is purely that of its starch. The chief exporting countries are Malaysia and Indonesia, but the palm is also used in neighbouring countries such as Thailand and the Philippines.

Other Sago-producing Plants. Sago is also produced in a similar way from a number of other plants, though the quality of the product from some of them is held to be inferior. Most of these plants are other palms with a starchy pith. Among the more important of such palms are *Caryota urens* in India and Malaysia, *Metroxylon rumphii* in Indonesia, *Phoenix acaulis* in India and Burma, *Arenga saccharifera* (see also below) in south-east Asia, and the cabbage palm *Oreodoxa oleracea* in the American tropics. Sago is also obtained from the pith of two cycads, plants which although somewhat palm-like in appearance belong to a very different botanical group; these are *Cycas circinalis* in Ceylon and India, and *Cycas revoluta* in Japan.

2 **Sugar Palm** (*Arenga saccharifera*) grows wild in Malaysia and Indonesia and is sometimes planted in those countries. To make sugar, the tree must be tapped to obtain the sugary sap. The immature male spathe (whose spadix is shown on p. 17) is first beaten daily with a wooden mallet for about a fortnight to stimulate the flow of sap. As soon as the flowers open, the spathe is cut off at the base of the inflorescence and a receptacle attached to the wound to collect the sap. Cutting off a further slice of the spathe, beating, and replacement of the receptacle are repeated daily till the flow of sap is exhausted, by which time 3 pints per day may have been obtained for 7 weeks. Sugar is made by boiling the sap before it has time to ferment, until evaporation turns it into a thick syrup. On cooling, this solidifies into a thick brown toffee-like sugar.

Other Sugar-producing Palms. Sugar can be prepared in the same way after tapping the inflorescence or sometimes the terminal bud of many members of the palm family. Some palms which are mainly grown for other purposes are occasionally tapped for sugar, though this is rare if a full yield of fruit is wanted as tapping naturally weakens the palm. However the practice is found to a varying degree in different countries with the coconut palm (p. 19), the date palm (p. 107), the Borassus palm (p. 17), and *Caryota urens* (*see* col. 1 on this page, alongside).

There are other palms whose main economic use is to provide sugar. One rather prolific yielder is the nipa palm (*Nipa fruticans*), a small palm which grows wild in the brackish water of estuaries in south-east Asia. The wild date palm as a sugar producer is described on p. 17.

TREES SMALL SCALE   **SPADICES × ⅛**

1 SAGO PALM   1A Part of flowering spadix
2 SUGAR PALM   2A Flowering female spadix

SEAWEEDS

Many seaweeds are used for food, in various parts of the world, either directly, or indirectly in the form of extracts or chemical derivatives. The following are only a small selection.

1 **Laver** (*Porphyra umbilicalis*) is used for food in south Wales, Ireland, and some other European countries. It is usually washed and boiled before being sold and at this stage it looks rather like a spinach purée, except that it is dark brown in colour. In south Wales, it is a traditional breakfast food, coated in oatmeal before being fried and served with bacon and eggs. It is also eaten with potatoes and butter.

Laver is common on rocks and stones all round the coasts of the British Isles and other temperate North Atlantic countries. It is one of the Red Algae (Rhodophyta) and is rosy-purple, turning to olive-green or brown. The thin, flexible, wavy-edged frond is attached to the rock on which it grows by a small, disc-shaped holdfast.

The genus *Porphyra* has a world-wide distribution and several species are eaten, especially in China and Japan. The Japanese cultivate *Porphyra* by sinking bundles of bamboo shoots offshore. When a good crop of the young seaweed has become established, the bundles are transferred to less salty estuaries, where they will grow better than in undiluted sea water. The Japanese and Chinese use *Porphyra* in many ways, in soups and stews, as a covering around balls of boiled rice, and in pickles, preserves, and sweetmeats. It is said to have a high protein content and to be rich in vitamins B and C.

2 **Dulse** (*Rhodymenia palmata*) is sometimes eaten fresh in salads, or it is washed and dried for later use, either cooked, or as a masticatory like chewing gum. In New England, the dried seaweed is eaten as a relish. In Kamchatka, it is used for making an alcoholic beverage. Although one of the best known of our edible seaweeds, Dulse has little to recommend it — it is tough and has little flavour other than that of salt.

Like the previous genus, it is a Red Alga (Rhodophyta). The dark red, rather leathery, wedge-shaped frond is usually divided dichotomously or palmately, and old fronds often have rows of smaller, thinner 'leaflets' along their margins.

3 **Carrageen** (*Chondrus crispus*), also known as 'Irish Moss', is used in the form of an extract (carrageenin), in table jellies, blancmange, ice cream, salad dressings, confectioneries, soups, etc., as an emulsifier. It is also used medicinally, for the treatment of coughs. The United States is one of the principal commercial sources, but carrageen is widespread on temperate North Atlantic Coasts. Brittany and Ireland are important sources on this side of the Atlantic.

Carrageen is another Red Alga (Rhodophyta). Its fronds are attached to rocks by a disc-like holdfast and have a 'stalk' of varying length, branching repeatedly into a series of flattened segments, more or less fan-like in general appearance.

4 **Knotted Wrack** (*Ascophyllum nodosum*) is a common species on sheltered beaches around the British Isles and other temperate Atlantic countries. It is one of the Brown Algae (Phaeophyta), a group of plants which are generally less palatable to man in their natural state than the edible Red Algae, but which are used in the form of seaweed meal, mixed with other meals, as food for livestock. Together with Oarweed (*Laminaria* spp.), Knotted Wrack is an important source of alginates. These are organic substances which are used for thickening soups, emulsifying ice-cream and soft drinks, and gelling confectionery, jellies, puddings, etc. They can also be made to form thin films, which are used as edible sausage 'skins'. Knotted Wrack resembles some species of Fucus, such as Bladder Wrack (*Fucus vesiculosus*), in having branching fronds bearing conspicuous bladders, but its fronds have no midrib and its bladders are more variable in size and shape, sometimes as large as a walnut and often oval rather than spherical.

Oarweed (*Laminaria digitata*) is one of the best known seaweeds, common on the north temperate Atlantic coasts of Europe and North America. Its young stems have been used for food in Scotland and Ireland and the whole plants, boiled, have been used as cattle food in Scandinavia. Like the previous species, Oarweed is most important as a source of alginates, the chief producing countries being Britain (especially Scotland) and the U.S.A.

Other species of *Laminaria* have many culinary uses, in China, Japan, and elsewhere.

The genus *Laminaria* is easily recognisable by its very large, broad fronds, which, in *Laminaria digitata*, have a digitately branched blade and a cylindrical stem or 'stipe', 1 to 5 feet long, attached at the base by a much-branched holdfast.

LIFE SIZE

1 LAVER   2 DULSE   3 CARRAGEEN
4 KNOTTED WRACK

MUSHROOMS, TRUFFLES AND OTHER EDIBLE FUNGI

Many 'toadstools' are edible but some are highly toxic, even fatal. The most dangerous are those which bear some resemblance to edible species. With proper caution, we could make greater use of the wild harvest of edible fungi, of which the following are a few of the better-known.

1 **Truffle** (*Tuber aestivum*). This fungus, which grows underground in woods, especially beechwoods, is irregularly globose, 1 to 4 inches across, dark brown and warty, its flesh permanently solid, white, soon turning buff, with a network of white veins. This, the best flavoured of British truffles, is regarded as inferior to the French Périgord Truffle (*T. melanosporum*) which is used in *paté de foie gras*.

2 **Chanterelle** (*Cantharellus cibarius*). Common in woods, in summer and autumn, Chanterelles are rather firm-fleshed and need cooking longer than mushrooms. The funnel-shaped cap is egg-yellow, with paler flesh, having a faint odour reminiscent of apricots. The pale pinkish-buff spores are produced in narrow folds.

3 **Morel** (*Morchella esculenta*) grows in spring, often in woodland clearings. Morels have a distinctive appearance, their caps criss-crossed with irregular, pale brown ridges between which are darker brown hollows in which the spores are produced. The stalk is whitish becoming yellowish or reddish when old. There are several other British species of *Morchella*, all edible.

4 **Field Mushroom** (*Agaricus campestris*). The only 'wild' fungi which are commonly eaten in this country belong to the genus *Agaricus* and this species is usually regarded as being one of the best in flavour. Found in meadows and pastures in summer and autumn, it has a white cap, when young connected to the stem by a membrane (partial veil), which tears as the cap expands, its remains persisting for a time as a narrow ring around the stem. The gills are white at first, soon turning pink and finally dark purplish-brown. The spores are dark brown.
Agaricus bisporus, usually found on old dung, is distinguished microscopically from *A. campestris* by its spores being produced in pairs instead of in fours. The cap is generally pale brown. The cultivated mushroom is a white form of this species, traditionally grown on beds of stable-manure, although other composts have been tried with varying degrees of success.

5 **Blewits** (*Lepista* species) have a pleasant smell and are good to eat, cooked like mushrooms. *Lepista saeva* is found in open grassland, in autumn. It is mushroom-shaped, with a greyish or brownish cap, tinged with lilac or purple, white-fleshed when young. The stem is whitish with bluish streaks. The Wood Blewit, *Lepista nudum*, grows mainly in woods in late autumn and is wholly lilac or purple.

6 **Oyster Mushroom** (*Pleurotus ostreatus*) is rather flavourless, but as it often grows in colonies it is easy to collect and it dries and keeps well. The colour of the cap varies with age, from dark bluish-grey to pale brown. The widely spaced, yellowish-white gills merge into the very short stem which attaches the side of the cap to the host tree — often a beech. This fungus causes heart rot in the timber of trees; it also grows on stumps, fence posts, etc.

7 **Cep** (*Boletus edulis*). A member of a large genus, most (but not all) of which are edible, this fungus is widely eaten on the Continent, fried when fresh, or dried and used in casseroles and soups. It is common in woods (especially beech woods) in summer and autumn. It has a brown, smooth, moist, shining cap, its flesh white, often tinged with pink. Beneath is a spongy mass of vertical tubes, white at first, becoming yellowish-green, in which the brown spores are produced. The stalk is stout, pale brown, with a fine network of raised, white veins towards the top.

8 **Shaggy Parasol** (*Lepiota rhacodes*) is a member of a large genus with more than 50 British species, many of them edible. The Shaggy Parasol has a cap 3 to 7 inches across, covered with large, yellowish or brownish scales, except for a smooth brown disc in the centre. The long, stout, whitish stem is smooth. The flesh is white, turning red when cut. It grows in open spaces in woods and gardens, usually in rich soil.

Parasol Mushroom (*Lepiota procera*), also highly esteemed, has a distinct dome in the centre of the cap and a brown-mottled stem. It grows in grassy places, often near trees.

9 **Fairy-ring Champignon** (*Marasmius oreades*) forms the well-known 'fairy-rings' on lawns and short-turfed pastures. Its cap is brownish, often tinged with pink, paler when dried, slightly domed in the centre. The gills and the slender tough stem are pale buff. The flesh is white and has a pleasant, mushroom-like flavour but the tough stem should be discarded. This fungus is easy to dry and keeps its flavour well when reconstituted by soaking. It can also be pickled and used for making a ketchup.

10 **Giant Puff-ball** (*Calvatia gigantea*) grows in woodland and pastures, often to the size of a football and, exceptionally, up to 5 feet across. It can be eaten only when young, white, and firm-fleshed. With age, it turns yellowish and then brown. A mature puff-ball contains an astronomical number of powdery, olive-brown spores. There are several other edible Puff-balls, but none as large as this species.

TWO-THIRDS LIFE SIZE

1 TRUFFLE	2 CHANTARELLE	3 MOREL
4 FIELD MUSHROOM	5 BLEWIT	6 OYSTER MUSHROOM
7 CEP	8 SHAGGY PARASOL	
9 FAIRY-RING CHAMPIGNON	10 GIANT PUFF-BALL	

SOME BRITISH WILD PLANTS

Primitive man relied on wild plants for his food and even in technologically advanced countries some wild plants are still utilized today. In Britain, more than 300 wild plants have been described as 'edible', although some have been used only in times of poverty or famine. Many have been taken into cultivation. The following British wild plants (native or naturalized) are described elsewhere in this book: Hazelnut (p. 27), Wild Crab (p. 47), Rose (p. 63), Sloe (p. 67), Blackberry (p. 79), Cloudberry (p. 79), Dewberry (p. 79), Bilberry (p. 83), Cranberry (p. 83), Strawberry Tree (p. 83), Chicory (p. 111), Dandelion (p. 111), Juniper (p. 137), Hop (p. 137), Wormwood (p.137), Caraway (p. 139), Fennel (p. 139), Peppermint (p. 141), Spearmint (p. 141), Marjoram (p. 141), Tansy (p. 145), Chamomile (p. 145), Samphire (p. 147), Sweet Cicely (p. 147), Lovage (p. 147), Mustard (p. 153), Watercress (p. 153), Rocket (p. 153), Asparagus (p. 163), Sea Kale (p. 163), Chives (p. 167).

1 Elder (*Sambucus nigra*). Elderberry wine, made by fermenting the juice extracted from the ripe fruits, is well known. It is a rich, dark purplish-red wine, looking something like port, with which it has sometimes been blended, fraudulently. Elderflower wine, still or sparkling, is made from the corollas of the flowers, shaken off the inflorescence and carefully separated from stalks and other parts of the plant which have purgative properties. It has a very distinctive flavour and bouquet. The elder contributes to two notable conserves — elderberry and apple jelly, and gooseberry and elderflower jelly.
Elder belongs to the Honeysuckle family (Caprifoliaceae). It is a deciduous shrub or small tree, common in woods, hedgerows, and waste places. Its branches have brownish-grey, corky bark. They contain a large proportion of soft, light, whitish pith — used by botanists for holding flexible parts of plants which are to be hand-sectioned with a razor. The creamy-white flowers are bisexual. The fruit is green at first, and normally purplish-black, rarely greenish-white, when ripe.

2 Barberry (*Berberis vulgaris*) was formerly much planted for its edible fruits, as well as for 'ornament', but today it has to compete, in gardens, with many other introduced species and hybrids. It is possibly native in a few places in England, in hedgerows, etc., but it has been eradicated in many localities because it is an intermediate host of Black Rust, a fungus which attacks cereals. The red berries are pleasantly acid and used to be made into jelly, candied in sugar, or pickled and used for garnishing.
The Barberry belongs to the family Berberidaceae. It is a spiny shrub, 3 to 7 feet high, with bisexual, yellow flowers, borne in a pendulous raceme, succeeded by bright red fruits.

3 Good King Henry (*Chenopodium bonus-henricus*) was formerly cultivated for use as a green vegetable (like spinach). Although rarely grown today, the plant has become naturalized in some localities, usually near old gardens or buildings.
Good King Henry is a perennial herb, belonging to the Goosefoot family (Chenopodiaceae). It is a native of Europe, West Asia, and North America. It grows 1 to 2 feet high and bears rather fleshy, triangular, arrow-head shaped leaves. It can be distinguished from other species (*Chenopodium* is a large genus) by its leaf-shape and by the way the long stigma sticks out of the small, green flower. Several other species of *Chenopodium* and of the related genus *Atriplex* have been used as substitutes for spinach.

4 Stinging Nettle (*Urtica dioica*) is an all-too-familiar weed with a surprising number of useful attributes. The young tops, gathered when about 6 inches high, can be used as a green vegetable, usually in the form of a purée like spinach, but their rather earthy flavour is not to everyone's taste. From Scotland comes a recipe for nettle pudding, made with leeks or onions, broccoli or cabbage, and rice, boiled in a muslin bag and served with butter or gravy. Nettle beer and nettle tea are old country drinks, and preparations of the plant are used in herbal medicines. Dried nettles can be fed to livestock, but few animals will eat the growing plants. In wartime, nettle fibre has been used in large quantities for making textiles.
Nettles belong to the family Urticaceae, and have stinging hairs. The stinging nettle is a perennial. Its lower leaves are longer than their stalks and the small, green, male and female flowers are borne on different plants. The less common small nettle (*Urtica urens*) is an annual. Its lower leaves are shorter than their stalks and the unisexual male and female flowers are borne on the same plant.

5 Sorrel (*Rumex acetosa*) is one of several allied species which are sometimes cultivated for use as green vegetables, like spinach, or in salads or sauces. It is a common plant in grassland and open places throughout the British Isles.
Sorrel belongs to the Knotweed family (Polygonaceae). It is a perennial herb, from a few inches to 3 feet in height, with acid-flavoured, arrow-head shaped leaves which have downward-pointing basal lobes. The male and female flowers are borne on different plants.

TWO-THIRDS LIFE SIZE
1 ELDER   2 BARBERRY
3 GOOD KING HENRY   4 STINGING NETTLE   5 SORREL

THE DOMESTICATION OF FOOD PLANTS

Early man, following the habits of his ape-like ancestors, first made use of food plants by gathering the edible parts of wild plants. Excavations of human remains from the Iron Age in Denmark show that the meals of these people included the seeds, not only of various grasses, but of such plants as corn spurrey, hemp-nettle and ribwort which we should not think of as food plants today. However, gathering from wild plants is still carried on, even in the highly civilised countries of western Europe when people go out to pick wild blackberries or mushrooms. In some parts of the world there are considerable trades in produce gathered from wild plants which have never been taken into cultivation, such as Brazil nuts in South America. In times of famine or food scarcity, people revert to making more use of wild plants which are not normally eaten, as when in Britain during two world wars in the twentieth century nettles were gathered and cooked for use as greens.

In time man came to realise that he could obtain a larger food supply by collecting seeds of the most convenient food plants and sowing and harvesting them in a plot where he could keep down weed competition. Some of our oldest crops, such as wheat and barley, seem to have come into cultivation in the Middle East between 8000 and 5000 B.C. In China, rice is recorded as already a staple crop in 2800 B.C. The adoption of crop cultivation enabled men to live in denser populations and was followed by the development of civilisations with some people living in towns in such areas as Mesopotamia and Egypt.

In these early fields, plants which had favourable characteristics for cultivation, such as good storage qualities, easy germination, and uniform ripening without the seeds becoming detached from the plant, must have selected themselves for further propagation. Man himself must also in the course of time have noticed that there was genetic variation between plants, and it must have occurred to some farmers at least, to select the seed of outstanding plants for re-sowing. It is thought that some later domesticated crops, such as oats and rye, were first noticed by man growing as weeds in the crops first favoured; their seeds would occur as impurities in harvested wheat or barley grain, and in conditions that specially suited them there would be natural selection towards them, aided eventually by conscious separation by man. As time proceeded, and man's selection of crop seeds became more deliberate and more refined, the cutlivated forms of many crops became more and more divergent from the wild ancestors from which they had been derived. With the progress of civilisation, increasingly intelligent attention was given to these matters. By the eighteenth century in Europe, seed selection had become a fine art in the hands of such men as the Vilmorin family in France, and the yields of many crops had been raised beyond what was possible in medieval times. From the late nineteenth century onwards, this process was greatly speeded up by the development of the modern science of genetics. Now man began not only to select parent plants, but to make hybrids between species and varieties, to search the world for plants with promising characters, and even himself to create new genetic variations by treating plants with chemical substances or bombarding them with X-rays. The effect has been to change the character of some crop plants to an extent where our ancestors would hardly recognise them, and in many countries to increase the average yields of important crops by twice or more during the past century.

The result of this prolonged action of man upon his food crops is that the food crops now grown can be grouped into a number of classes in regard to their relationship to their wild ancestors. These classes are as follows (only a few examples are quoted in each class, as the origins of many individual food crops have been treated earlier in this book):

(1) The cultivated plant is still practically identical with its wild ancestor. Examples among plants native to Britain are seakale and water-cress. These are usually minor crops, already reasonably satisfactory as a food source, providing little incentive for improvement.

(2) The crop has been considerably improved but is still recognisably the same species as its wild parent. The cultivated carrot and parsnip differ from their wild ancestors, still common in Europe, mainly in selection for increased size and tenderness of root. The garden pear-tree differs chiefly in better size and taste of its fruit from its parent form which grows wild in western Asia.

(3) The cultivated plant has arisen from natural hybrids between two or more wild species, which were selected by man as useful for his purposes. Thus the cultivated bananas are all derived from hybridisation between two wild species *Musa acuminata* and *Musa balbisiana* which still exist; but the cultivated differ from the wild in that they have sterile seedless fruits and would not survive unless continuously planted by man.

(4) Some crop plants, being hybrids of species which were already in cultivation, originated by chance and some were deliberately created by man. Our modern wheats appear to be derived from a primitive cultivated wheat, itself the product of a cross between two wild grasses (*Triticum boeoticum* and *Aegilops speltoides*), which gained further valuable characteristics by crossing with another weed grass, *Aegilops squarrosa*. Rather similarly, the kinds of maize first cultivated or collected by man seem to have hybridised with species of *Tripsacum* and with teosinte (*Euchlaena mexicana*) to produce more recent forms. There are many examples of deliberate hybridisation by man among the citrus fruits (*see* p. 89).

(5) Some cultivated plants are clearly sufficiently different from any wild plant to be classified as separate species. More or less closely related wild plants usually exist, but it is often impossible to tell whether the cultivated plant is derived from them by variation or hybridisation, or from an ancestor which is now extinct. The wild ancestors of the broad bean, lentil, onion, ginger, and date palm are thought to be extinct.

It must not be thought from the foregoing paragraphs that the origin of all food crops is necessarily ancient. Some are of quite recent development. An outstanding example is sugar beet. Selection of beet plants for a high sugar content was only thought of in the eighteenth century, and the increase by many times of the yield of sugar obtainable from this crop since that time is one of the triumphs of plant breeding. The cultivated strawberry of today is derived from a chance cross between two wild species from America in France in 1790, and the loganberry from a cross between a blackberry and a raspberry in America in 1881. Other food plants may perhaps be domesticated in the future. Yeast fungi cultured on a suitable medium have already been processed for human food, and could help to solve nutritional problems in some parts of the world. Single-celled algae grown in nutrient solutions have been shown experimentally to be capable of giving a very high yield of food from a small area. Modern techniques of extracting protein from the leaves of plants for direct human consumption indicate that there are many plants which could be exploited in this way in the future.

Our existing food plants originated, as we have seen, in many different parts of the world. The Russian botanist Vavilov, who devoted much research into their origins, concluded that there were up to 12 main centres of origin in different regions, though there is no doubt that some food plants also originated in other scattered places. The majority of these centres of origin were in Asia, Africa, and the Mediterranean basin, though there were at least three in America, where the development of a range of food crops different from those of the Old World proceeded entirely independently until Columbus reached America. The centres of origin are mostly situated in mountainous tropical or sub-tropical regions, environments which in Vavilov's view were particularly suited for a diversity of types of flowering plants to evolve. The story of how our food plants spread outwards from their centres of origin, and particularly of the great interchange of plants between the two hemispheres following the discoveries of Columbus, is told in the next article. Man's exploitation of plants to provide him with food has extended not only to many geographical areas but to many different families of the plant kingdom, and it is this subject which now falls to be discussed in the last section of this article.

The Botanical Affiliations of Food Plants

Very little of our food supply is derived from non-flowering plants. Mushrooms and truffles, a few seaweeds, the seeds of some pines, and some cycads which are a minor source of sago show how slight and unimportant are these sources of human food. The angiosperms or flowering plants are therefore our chief source of plant food. Within them, major food plants can be found amongst both dicotyledons and monocotyledons, and are scattered through a very wide range of botanical families. Two families however stand out as particularly important and were independently exploited by man to provide many of his chief foods both in the Old World and the New. The first of these is the Gramineae or grass family, to which belong all our cereal crops and sugar-cane, and from which we get a few other food products such as bamboo shoots. The cereal grains provide a convenient source of carbohydrate and have also a fair protein content, and the vast majority of mankind has always relied on a cereal as the basic staple of diet, with wheat and rice, and in America maize, as the most important examples. The sugar-cane provides more calories of human food per acre than any other crop, but is useless except as a source of carbohydrate. The other family on which man drew largely in both hemispheres is the Leguminosae, providing pulse grains of the pea and bean type, and valuable for the notably high protein content of their seeds associated with the ability of these plants to assimilate atmospheric nitrogen through the bacterial nodules on their roots.

No other families provide anything like such an important range of food crops as the grains and the legumes, but some are of distinct importance in particular climatic regions. In the temperate zone, the Rosaceae include a notable group of fruits (apple, pear, plum, cherry, apricot, almond, peach and a number of wild and cultivated berries), but only one member of this family, the loquat, is at home in subtropical latitudes. The Cruciferae are again a family primarily of the temperate zone including a wide range of vegetables of the cabbage, turnip, and cress groups, and also rapeseed and radish. In the tropics, the most important family after the Gramineae and Leguminosae is probably the monocotyledonous family of the Palmae, as this includes the oil, coconut, date, sago, and sugar palms; the oil palm is notable as yielding more edible oil per acre than any other crop.

Apart from these families, it is difficult to pick out others which are of special importance as food sources. The Solanaceae include a remarkable diversity of food crops in potatoes, tomatoes, and red peppers. The Rutaceae include in the single genus *Citrus* a large number of fruit crops, as do the Anonaceae, though less importantly, in the genus *Anona*. But so scattered through the range of flowering plants are the remainder of our food crops that many of the most important ones are the only species of their family cultivated for human food. Notable examples of this are the grape-vine in the family Vitaceae; the olive in the Oleaceae; and the sweet potato in the Convulvulaceae.

THE SPREAD OF FOOD CROPS AROUND THE WORLD

In prehistoric times food crops, after their first domestication, spread slowly outwards from their centres of origin to such other areas as were easily accessible by land travel. Thus before written history began, even such outlying areas of western Europe as Britain had received cereal crops like wheat and barley which originated in western Asia. Field beans and peas, originating in Asia and the Mediterranean, have been discovered in excavations of Swiss lake dwellings of the Bronze Age. But the number of crops then being grown in Europe was much smaller than it is now. The development of food crops in the Americas and in the Old World was entirely independent. When the first Europeans explored America, they found that the only food crops common to both hemispheres were the coconut palm, which was only being grown in a few localities in America, and some gourds of minor importance.

The crucial date for food crop distribution is thus 1492, the year in which Columbus first sailed to America and made possible the great interchange of crop plants between the two hemispheres which took place during the following centuries and in effect doubled the vegetable resources of both sections of mankind. This process was begun by Columbus himself, who brought back maize (corn) on his very first voyage from the West Indies to Europe, and on his second voyage took with him seeds of European crops to America. But to understand the significance of the interchange which followed, it is necessary to consider each continent separately, the food plants which it had originated or received by 1492, and those which it received in post-Columbian times.

Europe

European agriculture did not stand still between prehistoric times and the great voyages of discovery. The Romans extended to many parts of their empire some crop plants of Mediterranean origin, including perhaps peas, oats, and rye in some areas. The Arab invaders of Spain introduced, or at least first popularised, in southern Europe a number of food plants with which they were familiar including rice, sugar-cane, sorghum, and some of the citrus fruits. New crops were still spreading in the fifteenth century whose last decade saw the voyages of discovery made: buckwheat was still new in some regions of France; hops were extending slowly through northern Europe and were not cultivated in England till the early sixteenth century.

The consequence of the voyages of Columbus was that the first American crops to reach Europe were introduced through Spain, though they then spread fairly quickly to other European countries as well as to North Africa and the nearer parts of Asia. Two of these food crops became particularly important in the dietary of Europe. Maize by the end of the seventeenth century was a staple food crop in northern Spain and Portugal, and only a little later in northern Italy, though in the Balkan countries which later became the chief maize-growing region in Europe, cultivation did not become important till the eighteenth and nineteenth centuries. The potato, already being sold in Spain in 1573 and introduced independently to the British Isles by their own seamen, was more quickly adopted into general cultivation in Ireland than any other American crop in Europe. This crop, however, took a long time to reach the outlying parts of Europe and was only introduced into Russia and Scandinavia in the eighteenth century. Among food plants of lesser importance introduced to Europe through Spain were tomatoes, chillie peppers, and sweet potatoes. The French explorers of Canada were responsible for the introduction to Europe of the Jerusalem artichoke and probably of the beans which we have ever since called 'French' beans.

By contrast with the voyages of Columbus, the opening up of sea communications with eastern Asia by the great voyage of Vasco da Gama in 1497 was not followed by many plant introductions, though rhubarb, already exploited in Asia as a food plant, was in course of time brought to plant in the gardens of Europe. Some food plants of nearer Asia were however at this period still being extended to new parts of Europe; thus we have records of the introduction to England of the

195

almond and apricot in 1548 and the peach in 1562. It may be said that by the eighteenth century all the food plants which are now of importance in Europe had already been introduced, even if not widely taken up. Subsequent introductions due to modern botanical exploration or the work of plant breeders have mostly been of vegetables or fruits which are of slight dietary importance, such as sweet corn and the asparagus pea.

Asia

Asia, the continent in which so many of man's food plants originated, has had less need than any other to supplement its food sources with foreign plants, and to this day has done so to the least degree. Some American plants nevertheless reached it quite early in post-Columbian times. Amongst the routes by which this could happen were Venetian trade with the Levant, Portuguese voyages to India, and Spanish voyages from Mexico to their Philippine colony. Maize had certainly reached the Euphrates by 1574. Amongst early records of American food plants being cultivated in India are pineapples in 1583, papaya in 1600, and sweet potatoes in 1616. The introduction of chillies from America enabled Indians to make hotter curries than had been possible before. But some American food plants took a surprisingly long time to reach Asia. Cassava (tapioca) was first taken to an Asian country, Ceylon, in 1786. Potatoes seem only to have been introduced to India from England in the late eighteenth century.

Africa

While North Africa shared in the agricultural development of the Mediterranean basin, the number of food crops indigenous to Africa south of the Sahara is very limited; sorghum, millets, yams, cowpeas, sesame and oil palms are the most important. From perhaps about the beginning of the Christian era these were supplemented by the arrival of the banana, probably first brought by the Indonesian invaders of Madagascar. About ten centuries later, Persian voyagers are reputed to have brought some food plants, such as the mango and egg-plant, to the East African coast. But the number of available crops was still so small that when the American crops reached Africa, probably casually in the first place through the visits of ships engaged in the slave trade, many of them quickly passed into widespread cultivation. Among this group of crops were maize, cassava, sweet potatoes, groundnuts, and French beans. The new American crops were particularly valuable to the population of Africa in warding off famine, for two new root crops were added to the available ones resistant to locust damage, and a plot of cassava, which will keep for a long period in the ground, was soon recognised to be one of the best forms of famine reserve. Portuguese ships coming from Asia were probably responsible for the first introductions, to west Africa at least, of coconut palms and common rice. In later centuries all the food plants of the world have become available for appropriate climatic regions of Africa.

Australasia

The islands of the Pacific possessed as indigenous food plants the screw-pines (*Pandanus*) and perhaps the breadfruit. The Polynesians who colonised these islands from south-east Asia about the beginning of the Christian era brought with them in addition the coconut palm, yams, and taro. Whether the sweet potato, which is of South American origin, had already reached these islands before the era of European exploration has been a matter of argument. Such of these food plants as would grow in New Zealand were taken there by Polynesian immigrants culminating in the Maori settlement of the fourteenth century; the remaining food crops of New Zealand were added by European settlement in the nineteenth century. In Australia the aboriginal inhabitants cultivated no plants, and the introduction of all food crops dates from the European settlement in 1788.

America

The American Indians had already in prehistoric times evolved quite a satisfactory dietary from the plant resources of their own continent. This was based primarily upon maize (corn) as the staple cereal, supported by the root crops cassava and sweet potatoes and the legumes groundnuts (peanuts) and French beans. Such diets are still prevalent in much of tropical America today. The Spaniards early in the sixteenth century brought sugar-cane and bananas, two crops which now occupy large acreages in America. Other Old World introductions which in the course of time became important in tropical America included rice and the citrus fruits. The effect of the slave trade was to bring not only negroes but the food crops to which they were accustomed in West Africa, and such crops as yams, cowpeas, and pigeon peas have remained particularly important among negro populations in America and the West Indies. Crop introductions were, however, even more important to the temperate than the tropical parts of America, for the former had fewer indigenous crops. European settlement of temperate North and South America was based primarily upon the cultivation of their familiar European crops; wheat was and has remained a more important foodstuff than maize to the settlers and their descendants. Potatoes, although a crop of South American origin, were only introduced into the British colonies of North America from Ireland in the eighteenth century. Crop introductions have continued down to the present day and in the course of time Americans looked further afield than Europe and drew on Asia as well; a crop now so important in North American agriculture as the soybean was not introduced until the nineteenth century.

Modern Plant Exchanges

The foregoing paragraphs have described the rather haphazard way in which food plants were spread around the world in earlier centuries, culminating in the great post-Columbian dispersal after Europeans had reached America. At a slightly later period this dispersal became somewhat more scientific and made good use of the botanic gardens which were founded in various parts of the world during the eighteenth century. It was through the botanic gardens of Amsterdam and Paris that coffee, originating from a single plant from Java, was introduced early in that century to Dutch and French colonies in America, which later became the greatest coffee-growing continent. Captain Bligh, acting under British government orders, brought breadfruit plants from Tahiti to the botanic garden of St. Vincent in the West Indies in 1793, and at the same time introduced the first thick-stemmed or 'noble' sugar-canes to Jamaica. Tapioca first reached Asia through an introduction to the Mauritius botanic garden in 1736 and from that source came to the Calcutta botanic garden in 1794.

The age of simple introductions of species of food plants hitherto unknown in a particular region appears now to be practically over. It has been superseded, since the development of modern scientific plant breeding, by an exchange not of new species but of newly-improved varieties or strains of familiar species of food plants. The yield and quality of many food plants has been so constantly improved by plant breeders in recent times that all countries are anxious to obtain good new varieties as soon as they are developed; and, in this sense, the exchange of food plants between countries goes on more actively than ever. The enormous increase in average yields of wheat in Europe during the last century could not have taken place without exchange of varieties between the European countries in which breeders are active. The development of high-yielding hybrid maize in the United States is one of the agricultural triumphs of the twentieth century, and such maize helped to restore the food supplies of southern Europe after the Second World War. As agricultural science is increasingly applied in the developing countries, they too are beginning to benefit greatly from such exchanges. A very high-yielding rice hybrid, created in the Philippines in 1962 by crossing a rice from Formosa with one from Indonesia, has been widely distributed in Asia with very beneficial results. A wheat-breeding programme in Mexico which has produced wheats now giving high yields in sub-tropical countries as far away as India and Pakistan achieved success by crossing Mexican and Colombian wheats with dwarf varieties from the United States which were derived from crosses with a dwarf strain imported from Japan. In such ways, the exchange of food plants between nations seems likely to go on indefinitely with mutual benefit to all.

THE USES AND NUTRITIONAL VALUE
OF FOOD PLANTS

The Role of Food Plants in Human Nutrition

Before we discuss the nutritional value of foods of vegetable origin, it is necessary to remind ourselves that plants are not the only source of human food. A proportion of the diet is also provided by foods of animal origin; this proportion is usually higher in the economically advanced than in the less developed countries, though some nomadic stock-owning tribes are among the few people who eat more animal than vegetable food. Even many so-called 'vegetarians' eat animal foods in the form of milk, butter, cheese or eggs. A few strict vegetarians eat no animal foods at all, either from religious convictions as in the case of some high-caste Hindus or from personal choice in other cases. For such people it is desirable to eat a specially broad range of plant foods in order to have the best chance of achieving an adequate and balanced diet. But apart from them, any considerations about the value of plant food must be remembered as having a bearing only upon a part of the diet, though that part is particularly large and important in many tropical and sub-tropical countries.

Staple Foods

Frequently in the pages of this book, a food plant had been described as providing the 'staple' food of people in a particular region. What exactly does this mean? In earlier times, in nearly all communities there was a single foodstuff which was eaten in greater quantity than any other and formed the basic part of the diet, and even today when crop plants have been so widely exchanged and when in the richer countries diets have become so greatly and beneficially varied, a staple foodstuff can still be discerned among most human populations. The commonest staples are cereals, especially wheat and rice. Wheat products are the staple food of a very large part of the human race, especially those of European descent although they are also very important in the temperate parts of India and China. They are consumed in bread, biscuits, cakes, puddings, macaroni, breakfast cereals and other forms which together still make up a large part of the diet even in western Europe and North America. In older days, bread was often made of barley and rye so that these could have been described as staple foods, as could oatmeal in parts of Scotland; but the use of these cereals now hardly justifies this ranking anywhere. Rice, forming often a very high proportion of the diet, is the typical staple food of enormous human populations in Asia. Of the other cereals, maize (corn) is a staple food of large populations in South America, parts of eastern Europe, and in eastern and southern Africa; sorghum in some semi-arid areas of Africa and Asia; and finger millet and bulrush millet in other parts of those continents. After cereals, root crops are the most common kind of staple food; potatoes in Ireland, yams in some parts of West Africa, taro and sweet potatoes in the Pacific islands, are typical examples. It is rare to find crops other than cereals or root crops providing a staple food, but there are a few such cases: plantains (cooking bananas) among some East African tribes, and the banana-like 'ensete' in parts of Ethiopia are examples.

Nutrients obtained from Plant Foods

If we next consider the classes of nutrient which are required by the human body, and the extent to which they are available in different plant foods, we shall begin to see how far the great staple foods and the subsidiary foods which are eaten with them are able to provide a satisfactory diet. To do this, it is necessary to consider the different classes of nutrients separately:

(a) *Carbohydrates*. These are needed mainly to maintain body temperature and to provide energy for work and movement; their capacity to do so is measured in calories. The most important carbohydrate in this respect is starch, the main nutritional constituent of cereal grains and edible roots, which as we have seen are the common staple foods. The carbohydrates also include sugars. The most important of these are sucrose or cane-sugar, which is prepared in a nearly pure state from sugar-cane and sugar-beet; and sugars such as fructose which gives the sweet taste to so many fruits, and glucose which is found particularly in grapes and some vegetables. Since starch when eaten is converted into sugars in the normal process of digestion, it matters little nutritionally in which of these forms carbohydrates occur in food.

(b) *Oils and Fats.* These also are chiefly providers of energy, but being a more concentrated source of calories than the carbohydrates, they are particularly convenient in cold climates so that people can eat enough to keep themselves warm without taking in too great a bulk. Most plant foods contain at least a little oil, but prime sources are the oil and coconut palms, olives, and oilseeds such as groundnuts (peanuts), sesame, soya bean (soybean), rape, and others.

(c) *Proteins.* These are needed to build and repair body tissues, and are required in particularly large amounts by growing children. Proteins are obtained from both animal and vegetable foods. Their nutritional quality depends on the proportion they contain of the different amino-acids, of which some 20 occur in food materials but not more than 10 are essential in human nutrition. In this respect, animal proteins are a better-balanced food than plant proteins, and people eating a reasonable amount of animal protein are less in danger of suffering from deficiency of particular amino-acids. The plant proteins themselves vary in quality, cereal proteins for example being rather low in the amino-acid lysine, in which however the protein of legumes is richer. The chief sources of protein in plant foods are cereals, because of the large amount eaten, and leguminous grains which although richer in protein are eaten in lesser quantity; green leaf vegetables are also a protein source.

(d) *Minerals.* Although needed in much smaller amounts than carbohydrates or proteins, the human body requires appreciable amounts of salt, iron, calcium, and phosphorus, and smaller quantities of some other minerals, such as iodine to prevent goitre. Minerals again are largely obtained from animal foods in the diet, and salt is eaten as the mineral itself. Nearly all plant foods contain small amounts of minerals, and in a varied diet provide some of the supply; many green leaf vegetables are a notably rich source.

(e) *Vitamins.* These are substances which are only needed in very small amounts but are essential to health. Lack of particular vitamins is the cause of certain diseases such as scurvy, rickets, beri-beri and pellagra, which are fortunately less common than they used to be before their cause was understood. Vitamins are obtained from both plant and animal foods; plants do not contain vitamin A, but many provide the pigment carotene from which the human body can manufacture it; one vitamin, B_{12}, occurs only in foods of animal origin. Plants which are particularly rich sources of vitamins have been noted in the earlier text. The best insurance against vitamin deficiency is to eat a well varied diet including a range of fruits and vegetables.

Types of Diet and their Adequacy

Human populations can be placed in three main dietary groups, some members of each of which face nutritional problems which could be solved by different eating habits. The vast majority of mankind may be classed as basically grain-eaters, though the actual proportion of cereal in the diet may vary from a lower figure in the richer countries to such examples as maize providing 75 per cent of the calorie intake in parts of Rumania or 80–90 per cent in Malawi. Grain-eaters need not be short of calories unless the supply fails, and cereals contain enough protein to prevent gross protein deficiency among such populations, though neither its quantity nor quality are sufficient for perfect health, especially in infancy and childhood, unless protein from other sources is consumed. Certain classes of grain-eaters provide some of the classic examples of vitamin deficiency. Beri-beri disease due to deficiency of vitamin B_1, or thiamine, occurs chiefly among people who live too exclusively on rice; pellagra due to deficiency of nicotinic acid similarly among those who rely too solely on maize. Deficiency of vitamin A is also widespread in the tropics.

The second great dietary group contains those whose staple foods are root crops, with whom may be classed those who eat mainly plantains which are a nutritionally similar food source. The problem for these people is that their main foods provide an adequate supply of calories as starch but too little of most other nutrients. It is amongst such people that extreme cases of protein deficiency are generally found, especially in the form known as 'kwashiorkor', a deficiency of early childhood all too common in the wet tropics where root crops are chiefly relied on for food.

A third, and much smaller, dietary group of mankind eats principally animal foods and sometimes no plant food. This group includes some nomadic cattle-owning peoples in West and East Africa, some herdsmen in the central parts of Asia, and some Eskimos in the Arctic. Meat, milk, and in some cases blood are their main foods. Neither calories nor protein are usually lacking, but such a diet can bring on troubles such as constipation and rheumatoid arthritis and sometimes vitamin deficiencies, though these are often remedied where such peoples buy alcoholic liquors from others who cultivate crops.

Uses of Food Plants for Special Purposes

So far we have considered food plants as providing nutritionally essential ingredients in human diets, but it must be realised that numerous food plants are grown with other objects, chiefly to provide flavourings or spices or to serve for the preparation of attractive beverages. In neither case is nutritional quality in the product a prime object, though it is sometimes attained incidentally.

The various peppers, mustard, cloves, and nutmeg are typical food plants of the spice group. They are generally used in such small amounts that their nutrient properties are of little account, though the red peppers happen to be a rich source of vitamin C which prevents scurvy. Their great value is to make appetising the duller foods which people might otherwise be unwilling to eat. They were of great importance in medieval Europe where the diet during the winter months consisted largely of preserved fish or meat with little taste or an unpleasant one. They are currently of greater importance in the tropics where many foods are dull and tasteless, especially the root crops and plantains which are commonly eaten as a boiled mash; they may also have a stimulating effect on the digestive juices. Among flavourings which are not spices, mushrooms are a typical example having little nutrient value but contributing pleasantly to taste. Flavoursome foods seem likely always to remain in demand, and it has been said that long after we have discovered how to produce our basic nutrients synthetically, agriculture will still be required even if only to provide asparagus and strawberries and cream!

Another group is the plants used to make pleasant beverages. Tea and coffee have little nutrient value apart from the milk and sugar which may be taken with them, but are esteemed as mild stimulants. Cocoa has a more appreciable calorie content. Fresh or bottled fruit juices are not only especially appreciated as an alternative to water in hot climates where large quantities of liquid have to be drunk daily, but also contain useful amounts of vitamins and contribute some calories through their sugar content.

Fermented liquors are in a rather different class, being prepared chiefly for their alcohol content. Apart from Moslems and a few other groups who abstain from alcohol, most peoples of the world devote a part of their farm acreage to crops destined for fermentation or distillation. In Europe, grapes are grown for wine-making, and barley and hops for the manufacture of beer; whisky and gin are also made from grain crops. Rice wine is a popular drink in Asia. In the tropics, 'toddy' is the fermented product of various palms, local beers are made from various grains and from plantains, and 'arrack' and other distilled spirits derive from the same sources. Nutritionally, alcoholic drinks have an appreciable calorie content and in some parts of tropical Africa where the consumption of local beer per head may average a quart a day, this can provide 10 per cent of the calorie requirement. Amongst some populations near the margin of vitamin deficiency, such drinks can also be a useful source of vitamins, since these are not destroyed by any cooking process during preparation. Distilled spirits have a higher calorie content than merely fermented drinks, but less other nutritional value and there is less dietary justification for drinking them.

The Effect of Processing and Cooking on Nutritional Value

Considerable changes in nutrient value can take place in the produce of a food plant between its being harvested from field or garden and eaten by the consumer. In the case of cereals, milling removes as bran the outer layers of the grain and usually the embryo, and these contain nutrients in different proportions to the endosperm which remains. Differences in milling account for the different taste and texture of white and wholemeal breads. The problems which arise in rice milling are discussed on p. 8.

Problems also arise in the peeling of root vegetables. Although the peel of most of these must be removed before eating, and this is especially important in bitter cassava where the peel may be the most poisonous part, an excessive amount of food is wasted in some tropical countries by over-thick peeling of roots with crude knives. Potatoes contain more protein in the skin and outer layers than in the central flesh, and therefore not only more food but food of better quality is obtained by cooking them 'in their jackets'.

Some changes in food values occur simply by the passage of time during marketing or storage. The carbohydrate content of fruits and vegetables will decrease by respiration. Vitamin C is lost particularly quickly; in green leaf vegetables little of it may be left after two or three days. After 5 days, peas in pod had lost 24 per cent of their vitamin C, and potatoes 35 per cent. These losses are much smaller when foods are frozen, and vitamin C is also well preserved in canning.

The chief nutritional objects of cooking plant foods are to burst the starch grains and so make the starch more digestible, and to break down the cellulose cell walls so that the gastric juices can better reach the cell contents. Some vitamins, however, are partially destroyed by cooking. Potatoes when boiled may lose a quarter to a third of their vitamin C, of which they are an important source, and a quarter of their thiamine. Those fruits and vegetables which can be eaten raw, so that cooking losses are avoided, have a particular value in the diet, provided that they have been grown and marketed under hygienic conditions and are not carriers of disease germs.

INDEX